The Landscape of Pervasive Computing Standards

Synthesis Lectures on Mobile and Pervasive Computing

Editor

Mahadev Satyanarayanan, Carnegie Mellon University

Mobile computing and pervasive computing represent major evolutionary steps in distributed systems, a line of research and development that dates back to the mid-1970s. Although many basic principles of distributed system design continue to apply, four key constraints of mobility have forced the development of specialized techniques. These include unpredictable variation in network quality, lowered trust and robustness of mobile elements, limitations on local resources imposed by weight and size constraints, and concern for battery power consumption. Beyond mobile computing lies pervasive (or ubiquitous) computing, whose essence is the creation of environments saturated with computing and communication yet gracefully integrated with human users. A rich collection of topics lies at the intersections of mobile and pervasive computing with many other areas of computer science.

The Landscape of Pervasive Computing Standards
Sumi Helal
2010

A Practical Guide to Testing Wireless Smartphone Applications
Julian Harty
2009

Location Systems: An Introduction to the Technology Behind Location Awareness
Anthony LaMarca and Eyal de Lara
2008

Replicated Data Management for Mobile Computing
Douglas B. Terry
2008

Application Design for Wearable Computing
Dan Siewiorek, Asim Smailagic, and Thad Starner
2008

Controlling Energy Demand in Mobile Computing Systems
Carla Schlatter Ellis
2007

RFID Explained
Roy Want
2006

The Landscape of Pervasive Computing Standards
Sumi Helal

ISBN: 978-3-031-01352-2   paperback
ISBN: 978-3-031-02480-1   ebook

DOI 10.1007/978-3-031-02480-1

A Publication in the Springer series

SYNTHESIS LECTURES ON MOBILE AND PERVASIVE COMPUTING

Lecture #7

Series Editor: Mahadev Satyanarayanan, Carnegie Mellon University

Series ISSN

ISSN   1933-9011   print
ISSN   1933-902X   electronic

The Landscape of Pervasive Computing Standards

Sumi Helal
University of Florida

SYNTHESIS LECTURES ON MOBILE AND PERVASIVE COMPUTING #7

ABSTRACT

This lecture presents a first compendium of established and emerging standards in pervasive computing systems. The lecture explains the role of each of the covered standards and explains the relationship and interplay among them. Hopefully, the lecture will help piece together the various standards into a sensible and clear landscape. The lecture is a digest, reorganization, and a compilation of several short articles that have been published in the "Standards and Emerging Technologies" department of the *IEEE Pervasive Computing* magazine. The articles have been edited and shortened or expanded to provide the necessary focus and uniform coverage depth.

There are more standards and common practices in pervasive systems than the lecture could cover. However, systems perspective and programmability of pervasive spaces, which are the main foci of the lecture, set the scope and determined which standards should be included. The lecture explains what it means to program a pervasive space and introduces the new requirements brought about by pervasive computing. Among the standards the lecture covers are sensors and device standards, service-oriented device standards, service discovery and delivery standards, service gateway standards, and standards for universal interactions with pervasive spaces. In addition, the emerging sensor platform and domestic robots technologies are covered and their essential new roles explained. The lecture also briefly covers a set of standards that represents an ecosystem for the emerging pervasive healthcare industry.

Audiences who may benefit from this lecture include (1) academic and industrial researchers working on sensor-based, pervasive, or ubiquitous computing R&D; (2) system integrator consultants and firms, especially those concerned with integrating sensors, actuators, and devices to their enterprise and business systems; (3) device, smart chips, and sensor manufacturers; (4) government agencies; (5) the healthcare IT and pervasive health industries; and (6) other industries such as logistics, manufacturing, and the emerging smart grid and environment sustainability industries.

KEYWORDS

pervasive computing standards; pervasive computing middleware; programmable pervasive spaces; universal interaction in pervasive spaces; sensor platforms

Contents

Preface

This lecture presents some of the established or emerging standards and practices that are driving the technological evolution of pervasive computing systems. The lecture explains the role of each of the covered standards/practice and explains the relationship and interplay among them. Hopefully, the lecture will help piece together the various standards into a sensible and clear landscape.

The lecture covers four groups of standards and emerging technologies as shown in Figure P.1. The first group focuses on the basic elements, which are standards for sensors, actuators, devices, and appliances. Basic sensor representation, which is the first step in linking a physical entity into a cyber one is covered in Chapter 2 (Sensor Standards) and Chapter 3 (Service Oriented Device Architecture).

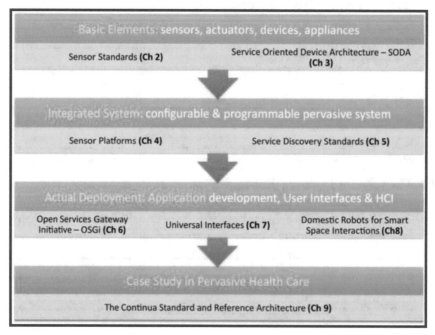

FIGURE P.1: Organization of the lecture.

The second group covers integration standards and emerging technology. Sensor platforms, the new emerging tiny computers, are presented in Chapter 4. Their role in integrating sensors and devices in a pervasive space is emphasized. Service discovery protocols are presnerd in Chapter 5 as a suite of integration standards that emphasizes the dynamic nature of basic elements in the pervasive space.

The third group looks at standards and technologies that facilitate programming and development in pervasive spaces. This includes service-oriented frameworks promoting service-oriented development of applications (Chapter 6), as well as universal user interfaces (Chapter 7) and emerging human–space interaction technologies focused in this lecture on domestic robots (Chapter 8).

Finally, Chapter 9 presents Continua—a vertical view and effort to pervasive standards, specifically in the domian of personal health.

The lecture is a digest, reorganization, and a compilation of several short articles that have been published in the "Standards and Emerging Tecnologies" department of the IEEE Pervasive Computing magazine (of which I am the department editor). The articles are edited to provide the necessary focus and uniform coverage in this lecture.

Audiences who may benefit from this lecture include: (1) academic and industrial researchers working on sensor-based, pervasive, or ubiquitous computing R&D; (2) system integrator consultants and firms, especially those concerened with integrating sensors, actuators, and devices to their enterprise and business systems; (3) device, smart chips, and sensor manufacturers; (4) government agencies; (5) the HealthCare IT industry; (6) as well as other industries such as logistics, manufacturing, and the emerging smart grid and environment sustainability industries.

I hope you enjoy the lecture, and hope to be able to extend it and keep it up-to-date over the years.

Dr. Sumi Helal
June, 2010

Acknowledgments

The author wishes to express thanks and gratitude to all authors who contributed to the IEEE Pervasive Computing magazine articles used in this lecture. The author and the series editor also wish to express their gratitude to the IEEE Pervasive Computing magazine and its editorial board for permitting reuse of some of the copyrighted materials.

I am particularly thankful to Shani Murray, IEEE Pervasive Lead Editor, for editing the original versions of several of the articles used in this lecture.

Eunju Kim, Ph.D. candidate at the Computer & Information Sciences & Engineering department at the University of Florida provided valuable support to the author by converting file formats and editing the Word templates and fixing the lecture format several times. Without Eunju's help this project would have taken a longer time, and for that, the author and the series editors are very thankful.

Dr. Sumi Helal
June, 2010

CHAPTER 1

Introduction*

Like it or not, tens of billions of lines of COBOL code are still in use today. Invented in 1959 by a group of computer professionals, COBOL empowered developers worldwide to program the mainframe and create applications still in existence today. Undoubtedly, COBOL owes much of its success to its standardization, which started with the American National Standard in 1968.

Yet these days, standards alone won't lead to success. With the invention of the PC and emergence of the network, we realized we also need new concepts and capabilities to program networks of computers. Standards such as TCP/IP and IEEE 802 played a major role in transforming the first computer network concept (Arpanet) to the Internet we know today. However, we also had to invent new computing models such as the client–server model, transactions, distributed objects, Web services, disconnected operation, and computing grids. Furthermore, we had to invent various middleware to support these emerging models, hiding the underlying system's complexity and presenting a more programmable view to software and application developers.

Today, with the advent of sensor networks and pinhead-size computers, we're moving much closer to realizing the vision of ubiquitous and pervasive computing. However, as we begin to create pervasive spaces, we must think ahead to consider how we'll program them—just as we successfully programmed the mainframe and, later on, the Internet.

1.1 LIMITATIONS OF INTEGRATED ENVIRONMENTS

Researchers have recently developed various pervasive computing systems and prototypes to demonstrate how this new paradigm benefits specific application domains (such as homeland security, smart homes for the elderly, habitat monitoring, entertainment, and education). In most cases, the researchers followed a system integration approach, interconnecting various physical elements and devices including sensors, actuators, microcontrollers, computers, and appliances using wired or wireless networks and connectors. Unfortunately, many of these systems and prototypes are rigid

*This chapter is based on the following contribution: Sumi Helal, "Programming Pervasive Spaces," the Standards, Tools and Emerging Technologies Department, IEEE Pervasive Computing magazine, vol. 4, no. 1, January–March 2005.

and inflexible; they lack scalability and are closed to third party participation. Furthermore, they have yet to demonstrate their ability to evolve and cope with change as new technologies emerge and as our understanding of a specific application area matures. We summarize the limitations of integrated environments below.

1.1.1 Nonscalable Integration

Any pervasive system is bound to consist of numerous heterogeneous elements that require integration. Unfortunately, the system integration task itself, which is mostly manual and ad hoc, usually lacks scalability. There's a learning curve associated with every element in terms of first understanding its characteristics and operations and then determining how best to configure and interconnect it. Also, every time you insert a new element into the space, there's the possibility of conflicts or uncertain behavior in the overall system. Thus, tedious, repeated testing is required, which further slows the integration process.

Even if an element has been previously integrated, there are no guarantees the integration effort is leverageable and the integration procedure is replicable. Consider a temperature sensor that needs to be connected to an embedded Java program to periodically report a refrigerated truck's temperature. Say you have two boxes—one from Hewlett-Packard containing a PDA running Linux, and another from Maxim containing a one-wire temperature sensor. These will require quite a bit of different physical and hardware interfacing. In addition, you'll need to write low-level software to interact with the sensor. Even worse, change the sensor, vendor, or the operating system, and you might have to completely rewrite the software and/or change the physical interface.

1.1.2 Closed-World Assumptions

Another problem is that an integrated environment is a relatively closed system—it's not particularly open to extensions or expansion, except perhaps accidentally. It's tightly coupled to a combination of technologies that happened to be available at development time. It's thus difficult—if not impossible—to add new technologies, sensors, and devices after system integration is complete and system has been deployed.

An integrated environment is also closed and restricted to only a few participants—the owners, designers, system integrators, and users. There's no easy way to let a third party participate. For example, an energy- and utility-efficient smart home developed in 2005 might not be compatible or interconnectable with a utility-saving sprinkler system developed in 2009 by a third-party vendor. Also, company B may not be able to provide maintenance and upgrade services to a system developed by company A. This will naturally lead to isloated, non-cooperative islands of smart environemnts—a "fragmentation" phenomenon.

1.1.3 Fixed-Point Concepts

Also problematic is the fact that our experience in building integrated environments is limited by the set of concepts we know at the time of development. This might sound like an always true statement regardless of whether we're doing pervasive, mobile, or distributed computing, but it's especially troubling in pervasive computing.

Take smart homes, for example. Unlike mobile phones, you can't upgrade and completely replace them every six months. Once built for a specific goal (such as to assist the elderly or individuals with special needs, save power, or support proactive health for an entire family), the home will likely be used for decades to come. That's why we need to ensure that our smart spaces will be "open-minded" about embracing not-yet-developed concepts. This might not be realistic, but smart pervasive spaces are bound to outlast any known set of pervasive computing concepts. Service gateways and context awareness are two examples of recent conceptual developments that have steeply influenced how we think of and build pervasive spaces. Surely other new concepts are on the horizon.

1.1.4 Lack of Programmability

Integrated environments are brittle systems that are difficult and expensive to change, customize, or personalize. They are not programmable unless a specialized API is designed and built into them. Such APIs would be expensive to build and suffer inherently from the *nonscalable integration* limitation previously discussed. Programmability in pervasive systems is a very important virtue to have because we rarely know *a priori* the exact way in which space elements should interact logically even though we often know what exact elements are needed in that space.

1.2 THE NEED FOR MIDDLEWARE AND STANDARDS

Moving beyond integrated environments will require a middleware that can automate integration tasks and ensure the pervasive space's openness and extensibility. The middleware must also enable programmers to develop applications dynamically without having to interact with the physical world of sensors, actuators, and devices directly. In other words, we need a middleware that can decouple programming and application development from physical-space construction and integration.

1.2.1 Self-Integration

Universal Plug and Play and other service discovery protocols are difficult to ignore when considering a middleware for pervasive computing. UPnP lets home computer owners connect devices to their PCs without having to manually integrate them (that is, without having to install drivers). We similarly need a standard or a middleware that let elements in a pervasive space integrate themselves automatically into that space. Such self-integration would lead to scalable, economical, and open pervasive computing—scalable and economical because we'd no longer need human system integrators (engineers or consultants working hundreds of hours and charging thousands of dollars), and open because third parties implementing sensors or devices could participate at any time in the pervasive space's life cycle.

Self-integration, however, requires a standard—which could be based in part on existing service discovery and delivery protocols such as UPnP and OSGi. The challenge is to find a standard whose adoption is possible by a broad category of players including sensor and device vendors and appliance manufacturers (Figure 1.1). It should be just as easy for, say, a heat sensor manufacturer to implement the standard as it would be for a plasma display manufacturer. If both the heat sensor and the plasma display were equally able to advertise their presence and register their services once brought into a space, we'd be much better off than we are now with integrated environments.

In reality, however, many existing sensors and other elements in a pervasive space can't participate in any standard or non-standard protocols. A heat sensor, for instance, doesn't have any processing or memory capabilities to engage in any protocols. How can such a sensor be self-integrated? It can't, at least not without a *sensor platform*—a hardware/software "adaptor" that the

FIGURE 1.1: Illustration of spaces labeled with compatible standards (e.g., UPnP and OSGi).

(heat) sensor could be instrumented with and physically connected to. Such sensor platform would provide two main functions—(1) establish the sensor as an input source to the platform, and (2) export a representation of the sensor and its readings and control to the rest of the world. Sensor platforms don't have to be powerful computers. They only need an embedded microcontroller, a small EEPROM memory and a communication and networking element.

Appliances are another challenge to the concept of self-integration. How can you integrate a floor lamp into a space? Again, you can't, unless you fit the floor lamp with a sensor platform. Similarly, sensor platforms are equally needed as middleware to self-intregrate actuators (e.g., switches, servo motors) and other complex devices (e.g., personal medical devices).

Sensor platforms may take on different form factors even though their functions remain essentially the same. One form factor is *embedded* in which the sensor or the device is already equiped with computing, memory, and communication hardware. In this case, the sensor or device needs only the software part of the middleware but no sensor platform hardware. Another form factor is *overlay*. For instance, most appliances use power plugs, and hence, invisibly integrating sensor platforms to power outlets would fit the appliance with a sensor platform the moment they are plugged into the power outlets (resulting in "smart plugs"). This way, we could integrate floor lamps, irons, microwaves, and the whole world of appliances into smart spaces, by simply plugging and powering up.

A handful of sensor platforms exists and are in use today even though only a few addresses self-integration as a primary goal. Sensor platforms are covered in detailed in Chapter 4.

1.2.2 Semantic Exploitation

Any middleware we use should also extend self-integration to include service semantics in addition to the service definition so that a joining entity could explore and fully participate in the space. For example, the temperature sensor (via its sensor platform) could offer information about its domain values (such as whether it measures in Celsius or Fahrenheit). It could also suggest other aggregated services that it could offer if and when other services become available in the space (such as a climate sensor if a humidity sensor is added). Exploiting semantics will go beyond integration and will let the pervasive space's functionality and behavior develop and evolve.

Space-specific ontologies will enable such exploitation of knowledge and semantics in pervasive computing. This again seems feasible and within reach. Ontologies for smart homes have started to emerge, so it shouldn't be too difficult to define other important ontologies such as for a classroom, coffee shop, bus station, commuter bus, train, or airport terminal, to mention just a few.

Overarching ontologies (non domain specific) could also be developed to addess device and crucial space aspects. For instance, user safety, space safety, and device safety could be addressed via semantic descriptions of what consistitute unsafe states for the user, space, and device, respectively. Take for example an elecric door opener, which is an actuator that triggers a mechanical strike mechanism. There is obviously a minimum time between successive actuations that must be observed. Opening and closing such door 50,000 per second (due to an error or a bug in the application) would most certainly damage the device. Describing the limitations on how a sensor or a device could be used could be part of such overarching ontology.

1.3   PROGRAMMABILITY

A critical goal for middleware is to present application developers with a programmable environment. In other words, the middleware should create and activate the functionality of an otherwise self-integrated yet *application-less* pervasive environment. If the middleware fulfills this role, it'll create a new paradigm in which the process of creating and integrating the physical world (sensor deployment phase) is separate from and decoupled from the process of designing and "engineering" the specific desired applications (programming phase). By comparison, an integrated environment has the applications integrated and bundled with everything else.

The middleware should let programmers perceive the smart space as a runtime environment and as a space-specific software library for use within a high-level language. For example, it should present all sensors and actuators in a form ready for use—perhaps as services. With special support to browse and learn such a dynamic library of services, a programmer should be able to immediately use the space sensors and actuators from within the application under development. Service composition would then be a natural methodology for developing applications on top of this middleware.

1.3.1   Service Orientation

Having a service view of every sensor, actuator, and device will enable rapid prototyping and a much faster development life cycle. For example, suppose you want an application that can control ambient lighting when the TV is on. A programmer quickly browses the space and identifies a room-light sensor service, a room-light dimmer service, a window-blind actuator service, and a TV appliance service. The programmer could then easily develop logic that uses all these services to sense the context of ambience and determine a possible action, which could in turn use additional services. So, an action might use the light-dimming actuator service and possibly the motorized-blind actuator service to bring light to ambient conditions.

The middleware's programmability aspect will not only empower application development but also support the notion of context-aware application development. Assuming a simple definition of context—"a particular combination of sensor states"—it should be straightforward for programmers to define contexts as special service compositions of relevant sensor services. Programmers could vary context sensors' properties to allow a variety of context production and consumption models.

1.3.2 Who Should Program a Pervasive Space?

Using middleware and standards, the role of engineers and system integrators diminishes. The paradigm shifts to IT professionals who are empowered to act as pervasive space programmers. *Pervasive Space Programmer* is not a commonly defined job by all means but will soon be. Such programmers will be highly productive and focused on the application's goals, since they do not have to worry about integration issues or interacting with system integrators and engineers. It should also be easy to train such programmers and thus to create a whole developer community for pervasive spaces based on standards and middleware.

However, we can't gain the benefits of pervasive computing without involving domain experts in application development. A psychiatrist, for instance, would be the best individual to program an at-home application to detect if insomnia is an experimental treatment's side effect. Similarly, a gastroenterologist would be the most appropriate person to test, nonintrusively, if an elderly patient's peptic ulcer is the result of an *H. pylori* bacteria caused by insufficient hygiene. Envisioning such scenarios helps reveal the need to change our programming and programmer models to accommodate domain experts as well as computer professionals.

1.3.3 Programming Models

Object discovery, reflection, and brokerage, all of which can deal with dynamic environments, have been useful mechanisms in object-oriented programming. Yet service discovery and lookup services have proven to be even more effective in such environments. We need similar concepts to effectively program pervasive spaces. Space reflection will be essential in providing application developers with a programming scope. As pointed out earlier, self-integration could provide all the information needed for space reflection in the simple form of service advertisements. Therefore, service-oriented programming seems more appropriate than the object model for pervasive computing. Service-oriented programming is also a much simpler model, which increases the chances of success in broadening the programmer model.

Indeed, we need a programming model that can deliver simple but powerful abstractions for a broad category of programmers. Some context-aware programming toolkits appear to fill this need [1]. A programmer presented with a context/condition/action model (similar to the event/condition/action model followed in active databases and other systems) might quickly develop better applications, but we need to further investigate this.

To achieve dependability and cope with behavior uncertainty, we need some "global" programming. Such programming won't be application-specific, but rather space-specific. It could be equivalent to administrative programming of distributed system tools and monitors. Any proposed programming model should address this need.

1.3.4   Integrated Development Environments

IDEs have revolutionized programmers' productivity and promoted the adoption of good software engineering practices. There's an even greater need for IDEs in pervasive computing. (Don't confuse IDE with IE—integrated environments or, as I call them, first-generation pervasive computing systems). Visual Studio, Forte, and Eclipse are great examples of IDEs.

Imagine a smart space being represented as an Eclipse-like project showing tabs for existing sensors, actuators, and contexts as well as service and context composition tools. Such an IDE would be initialized by pointing it to a space address (IP or URL) instead of to a project workspace directory. The IDE would use space reflection to initialize and to provide the programmer with the view and scope necessary for application development and debugging. The IDE could even be a graphical development environment in which services, context, and applications are represented or created graphically using a LabView-style building-block interface [2]. Developed applications might be committed back to the space as registered services. Without expanding on the range of capabilities these IDEs should offer, it should be obvious that such IDEs will contribute significantly to programming pervasive spaces. The main difference between a traditional IDE and a pervasive space IDE is that the former is a compile-time only environment (e.g., program editting, compiling, and debugging), whereas the latter requires run-time environment for proper development of applications.

· · · · ·

CHAPTER 2

Sensor and Device Standards*

In principle, the entire world can exploit sensors and networked sensor systems to great societal benefits. In practice, however, the domain of sensors is still a wild jungle in which nothing is quite standardized. To fix this, we need a device standard to characterize physical and electrical features and constraints, describe necessary data processing and message parsing procedures, and, more importantly as discussed in Chapter 1, enable automatic device integration and programmability into the world of computers we already have.

But standards that accommodate sensors and devices with diverse complexities and capabilities aren't easy to create or agree upon. Nevertheless, several standards and proposals are out there for examination and potential adoption. At the very least, we should consider them carefully and start paying more attention to the devices as first class citizens in the computing world.

In this chapter, we highlight five standards, some of which are labeled "device," "sensor," or "transducer," but all of them are equally capable of describing devices. If we consider sensors as simple or primitive devices, it follows that all five standards are also capable of describing sensors. We therefore make no distinction in this chapter between sensors and devices.

2.1   DIFFERENT STANDARDS

Let's first introduce the five emerging standards (Figure 2.1) briefly and then sift through them by discussing a variety of issues and examining their differences, similarities, and goals.

2.1.1   ECHONET

Originally initiated in Japan in 1998, the Energy Conservation and Homecare Network (ECHONET) standard specifies an open system architecture that enables the integration of a variety of home

*This chapter is based on the following contribution: Chao Chen and Sumi Helal, "Sifting through the Jungle of Sensor Standards," the Standards, Tools and Emerging Technologies Department, IEEE Pervasive Computing magazine, vol. 7, no. 4, October–December 2008.

FIGURE 2.1: An artist illustration of the wild jungle of devices and sensors and a number of emerging standard proposals.

appliances and sensors from multiple vendors [3]. It supports energy consumption monitoring and management and allows for networked applications and services that can access and control home appliances. To describe the interface of these devices, ECHONET provides a device specification that explicitly defines their properties and access methods. Therefore, to get a device approved and incorporated into the home network, vendors must create the same interface as specified in the standard to earn an ECHONET certification. ECHONET appliances are available primarily in the Japanese market.

Figure 2.2 shows a snippet of an ECHONET gas leak sensing device obtained from the ECHONET 1.0 specifications [4]. Three essential properties of the device are shown. Detection of threshold level of the sensor is the first property with a Set/Get access methods. Content of the property shows that for this sensor type, it is best to specify a range consisting of 8 values, with a minimum and maximum thresholds. The second property is checking gas leak occurrence status. It uses only Get access method (cannot be set) and specifies the return code for positive and nega-

Property name	EPC	Contents of property		Data type	Size	Access rule	Man-datory	Announcement at status change	Remark
		Value range (decimal notation)							
Detection threshold level	0×B0	Specifies detection threshold level in 8 steps		unsigned char	1 Byte	Set/Get			
		0×31~0×38							
Gas leak occurrence status	0×B1	Indicates gas leak occurrence status.		unsigned char	1 Byte	Get	O	O	
		Gas leak occurrence found = 0×41 Gas leak occurrence not found = 0×41							
Gas leak occurrence status resetting	0×BF	Resets gas leak occurrence status by setting 0×00.		unsigned char	1 Byte	Set			
		Reset = 0×00							

FIGURE 2.2: A snippet for a gas leak sensor in the ECHONET device object specification.

tive statuses. Status can be explicitly rest using the third property with a required reset value to be used.

2.1.2   IEEE 1451

The IEEE standard for smart transducer interfaces (IEEE 1451) describes a set of open and network-neutral interfaces for connecting and interacting with sensors and actuators through communication networks and processors [5]. The standard uses Transducer Electronic Data Sheet (TEDS) to specify information such as transducer identification, calibration, correction data, and measurement range. TEDS typically resides in the embedded memory of the sensor or actuator (transducer), which is an electrically erasable programmable read-only memory (EEPROM). Therefore, only transducers with embedded EEPROM can use TEDS. IEEE 1451 can be credited as the standard that digitized the morass of datasheets (papers or electronic) to a human/machine readable format. Moreover, it simplified the management and book keeping of these datasheets tremendously by storing them right inside the transducers themselves.

One objective of the IEEE 1451 specifications is to create a standardized interface that sensor or actuator manufacturers can follow to create cost-effective sensors and actuators that interface with a variety of networks, systems, and instruments. Numerous TEDS templates are available today. A manufacturer who produces accelorometer sensors for example must use the TEDS accelerometer template in order to get TEDS certificied. The templates require mandatory descriptive information but allow for vendor- and user-defined information to be included in the device TEDS sheet.

Figure 2.3 shows a screen shot of a TEDS editor supplied by Brüel and Kjær to its customers. The editor allows users to view and edit TEDS information directly from/to a transducer or through a file on the PC running the editor. The screen shot shows specifications of an acceleromter sensor by the same vendor.

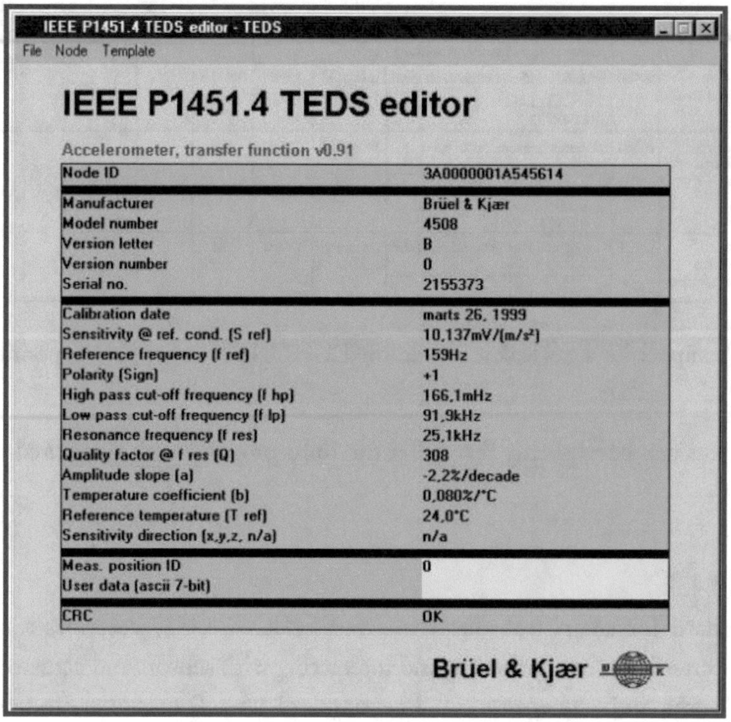

FIGURE 2.3: A snippet of a TEDS descriptor for an accelerometer.

The first section of the editor screen shows manufacturer, model, version, and serial number information, all of which are edited during production. The second section shows a mix of static and dynamic information about the acceloromter sensitivity and operational ranges and conditions, including calibration information (only last calibration is stored). The third section shows user-defined information such as the location of the measurement point where the acceloromter is used and any other uninterpretted user-specified data.

2.1.3   SensorML

The Sensor Model Language (SensorML) started as a funded NASA project led by researchers at the University of Alabama, Huntsville. In 2000, it was brought under the oversight of the Open Geospatial Consortium (OGC) where it served as a catalyst for the OGC Sensor Web Enablement (SWE) initiative [6].

There are several motivations behind SensorML, driven mostly by the OGC community. The need to preserve the process by which data are created is perhaps one of the strongest motivations behind SensorML. There is often a good distance between the input of a sensor (phenomenon) and an observation utilized by a scientist or a computer application. For instance, voltage is the output of a temperature sensor, which is processed through a voltage-to-temperature model, before it is adjusted through a calibration model, to then be recorded as an observation in meaningful units (Centigrade). In other words, SensorML can describe in detail the process by which an observation came about.

SensorML provides standard models and an XML encoding for describing all sensor processes. This includes the detection process (a physical process through which raw output of the sensor is generated), as well as one or more sensor models for measurement, filtering, conversion, aggregation, and other purposes.

Figure 2.4 shows a simple but complete example of a process model that uses two inputs, distance and time, to measure speed. The XML code defines each input and relates it to a unit of measurement (uom). It also defines the output and relates it to corresponding uom. The method of measurement (which would obtain speed as distance/time) is specified in the highlighted xlink: href link.

```
<sml:SensorML xmlns:sml="http://www.opengis.net/sensorML" xmlns:swe =
http://www.opengis.net/swe xmlns:xlink= "http://www.w3.org/1999/xlink"
xmlns:xsi="http://www.w3.org/2001/XMLSchema-instance"
version="1.0">

  <sml:ProcessModel id="PROCESS_ID">
     <sml:inputs>
        <sml:InputList>
           <sml:input name="distance">
              <swe:Quantity definition="urn:ogc:phenomenon:distance"
uom="m"/>
           </sml:input>
           <sml:input name="duration">
              <swe:Quantity definition="urn:ogc:phenomenon:duration"
uom="s"/>
           </sml:input>
        </sml:InputList>
     </sml:inputs>
     <sml:outputs>
        <sml:OutputList>
           <sml:output name="speed">
              <swe:Quantity definition="urn:ogc:def:phenomenon:speed" uom="m.s-
1"/>
           </sml:output>
        </sml:OutputList>
     </sml:outputs>
     <sml:method xlink:href="urn:uah:def:process:speedComputation:1.0"/>
  </sml:ProcessModel>

</sml:SensorML>
```

FIGURE 2.4: A SensorML process model of a speed observation.

2.1.4   DeviceKit

The DeviceKit [7] provides tooling and runtime capabilities enabling rapid prototyping and development of Java code to interface with hardware devices. It acts as an isolation layer between the hardware device and OSGi-based applications (OSGi is covered in Chapter 6). The DeviceKit has a platform-independent markup language-driven specification model where each device is described using the XML-based DeviceKit Markup Language (DKML). DKML allows the specification of hardware signals and protocols that are required for interfacing with a device, as shown in Figure 2.5. The DKML description is converted into appropriate Java code by the DeviceKit which automatically generates adapters for accessing and controlling the device as a software service.

The DeviceKit runtime abstracts a physical device into several software service layers: Connection Layer, Transport Layer and Device Layer, as shown in Figure 2.6. The layers are loosely coupled with mutually exclusive sets of functionalities thereby allowing an application developer to choose appropriate service options from each layer which can then be dynamically linked together at run-time using OSGi services mechanisms.

The Connection Layer is the lowest layer in the device abstraction hierarchy and is responsible for reading and writing byte streams of data to the hardware device over a specific connection interface.

The DeviceKit supports dynamic installation of connection adapters and provides native support for the following connection interfaces: IP, Bluetooth, File, URL, and RS-232 Serial Port.

The Transport Layer deals with the specific messaging requirements of the device. It is responsible for multiplexing byte streams received from the Connection Layer into messages and de-multiplexing messages into byte streams.

Integrating a device by converting it into a software service using the DeviceKit consists of a number of steps, some of which are automated whereas other require manual customization. The Device layer is automatically and fully generated from DKLM specifications that describe the device protocol and operational logic.

```
<signal id="LatitudeReport">
  <message id="LatitudeReportMessage">
   <bytes format="hex">FF,01,00,00</bytes>
    <filter id="filter0"><bytes format="hex">
       FF,FF,00,00</bytes></filter>
    <parameter type="short"><index> 2</index><size>
       2</size></parameter>
  </message>
</signal>
```

FIGURE 2.5: A snippet from a DKML device specification of the DeviceKit standard for a GPS device. The snippet describes the "LatidtudeReport" signal of the device's message exchange protocol.

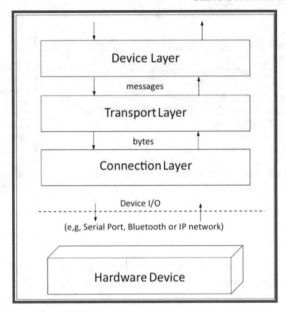

FIGURE 2.6: DeviceKit abstraction layers.

2.1.5   DDL

The Device Description Language (DDL) supports automatic device integration, including sensors and actuators [8–9]. Unlike the DeviceKit, DDL assumes devices have no networking capabilities and are connected to applications via sensor platforms such as the Mote [10] or the Atlas platform [11]. Thus, DDL follows a cross-layer design and only focuses on device interfacing.

A DDL descriptor file contains the information required for service registration and discovery, such as the name and model of the device and the description of its functions. It also describes the main operations of the device. DDL models a device from a data-oriented perspective, where each operation of a device is a collection of input/processing/output function chains. It describes the communications between a device and its corresponding service entity using "Signals." For example, a signal can be a voltage output from a pin, a number from an analog to digital converter (ADC), or a byte stream from a serial port, conforming to a certain well-known or proprietary protocol. Depending on the type of device, signals can be either unidirectional or bidirectional. "Readings" are the meaningful information that the service entity receives from the associated device. It usually specifies a conversion or a parsing method that converts raw signals to semantically meaningful data. For example, the analog temperature sensor in Figure 2.7 outputs a signal "0010000000," which is a 10-bit value from the ADC port. It is then converted to 325 degrees, a reading in Centigrade. In DDL, the combination of signals and readings together define the interface of a device.

```
<Sensor>
<Description>...</Description>
 <Interface>
  <Signal id="ADC1">...</Signal>
  <Reading id="Temp1">
  <Type>Physical</Type>
  <Measurement>Temperature </Measurement>
  <Unit>Centigrade</Unit>
  <Computation>
  <Type>Formula</Type>
  <Expression> Temp1 = (((ADC1/1023)
   * 3.3)-0.5)* (1000/10)</Expression>
  </Computation>
  </Reading>
 </Interface>
</Sensor>
```

A TMP36 Analog temperature sensor, manufactured by Analog Devices, Inc.

FIGURE 2.7: An analog temperature sensor and its DDL snippet (the snippet defines the sensor interface by showing how to convert an ADC signal into a temperature reading).

2.2   FROM PLAINTEXT TO MARKUP LANGUAGES

Many things jump to mind when people bring up the word "standard": a document written in some natural language, a set of formally defined procedures, a piece of computer software, and so on. All these options are actually employed in the real practice of defining sensor standards—for example, as part of the ECHONET standard, the ECHONET device object specification lists several classes, each one representing a device used in the home network. Each class' properties and methods are explained in plain English, with their types and value ranges explicitly specified in a way similar to a Java API document. The IEEE 1451 standard, on the other hand, uses a more succinct Interface Definition Language (IDL) to specify operation syntax, which reads like pseudo code and is programming language-independent. Although computer programs can't read either natural language or pseudo code, XML provides the additional machine readability. Thus, it's the common encoding basis for SensorML, DeviceKit, and DDL. This class of standards provides XML schemas to define language syntax, together with verbal descriptions that explain the semantics. For DeviceKit and DDL, open source language processors are available as supporting tools.

Although readability of sensor standards is largely enhanced by markup languages, it isn't the only issue of concern. Succinctness and scalability are equally important. A more succinct standard, for example, often leads to higher comprehensibility and an easier learning curve. Natural language is powerful and flexible, but without proper writing skills, it risks being verbose or ambiguous. Pseudo code is rather compact and tight, but the world lacks a standard of pseudocode syntax.

XML achieves a good balance between succinctness and accuracy by providing a well-formed and standardized language structure. It's also easily extendible by adding semantic constraints.

The ECHONET device specification is essentially a dictionary of devices, but maintaining such a repository by listing each and every member certainly isn't scalable. What if the repository expands? What if the device interface changes due to hardware upgrade? Each scenario would lead to modifications of the standard itself, which jeopardizes its stability. IDL-and XML-based standards, on the other hand, are more scalable—they've never attempted to standardize a specific device. By providing a generalized description tool, they let users freely define or modify any device and establish or expand their own device repositories.

2.3   BRIDGING THE PHYSICAL AND DIGITAL WORLD

It's no easy task to bring all the devices in a system together. Earlier developers of pervasive computing systems followed an ad hoc system integration approach, interconnecting various hardware pieces including sensors, actuators, appliances, and PCs with several network adapters and connectors. Unfortunately, many of these systems lack scalability and aren't flexible enough to accommodate new devices as novel hardware technology emerges. Recently, researchers and developers [12–13] have created several systems to demonstrate how Service-Oriented Architecture (SOA) enables easy and scalable integration of devices [14]. However, the assumption that SOA makes is that these interoperable services representing physical devices are already available—the question then becomes how to create a service entity within the SOA framework. To accomplish this, we must be familiar with the framework infrastructure and service interfaces. For example, a service bundle in the OSGi framework requires an interface class, an activator class, and an implementation class. So it might seem natural that this task falls onto a system integrator's shoulders. However, we could just as well argue that device manufacturers are in a better position to construct these services because they have the best knowledge of the device interface and the device's internal mechanisms, preferences, and constraints. There's a problem with this argument, however: although the great diversity of devices might overwhelm system integrators, a limited knowledge of SOA and software architecture in general could also challenge hardware manufacturers. The reality is that neither side would feel more comfortable—there's clearly a gap between the two.

Device standards could focus on this gap to close it. In addition to describing device hardware, the standard could prescribe a way to convert hardware pieces into software entities. These entities not only provide interoperability with other software entities but also allow interfacing (for access and control) with the corresponding devices. Depending on the software architecture proposed, the software entity representing a device could be a Java object, an OSGi bundle, or a Web service. DDL, for instance, follows SODA, which uses OSGi as its service framework. As part of its language processor, the DDL bundle generator converts a DDL device descriptor document to a service bundle running under OSGi. SensorML, on the other hand, doesn't use published software even though it proposed a big-picture architecture (sensor web enablement) that alluded to this necessity.

2.4   DIFFERENT SCOPES FOR DIFFERENT STANDARDS

To outline the full scope of sensor standards, let's consider a pinhead-sized analog temperature sensor. Tens of parameters appear on its data sheet: measurement range, accuracy, supply voltage, current, and operating environment, just to name a few. In addition, several performance diagrams and pin descriptions further characterize its electrical properties. An electrical engineer could consider these critical data, but are they of equal importance to an application programmer who only recognizes the sensor as a temperature service? How about system integrators? Do they need anything from this data sheet?

The goal of our discussion here is to compare available sensor standards to identify their descriptive scope. Figure 2.8 compares the five standards we investigated. From the physical world at the bottom to the digital world at the top, this figure shows how sensor standards bridge the gap.

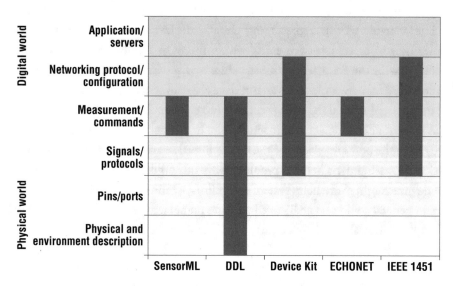

FIGURE 2.8: Five sensor standards. From the physical world at the bottom to the digital world at the top, we can see how sensor standards bridge the gap. The six layers labeled on the y-axis represent milestones along the way.

Six layers are labeled on the *y*-axis, each representing a milestone on this bridge:

- Physical and environment description explains a device's physical characteristics, such as form factor and operating environment. These external properties don't describe internal mechanisms or functions, and are usually of little interest to high-level application developers, but in many cases, information such as operating temperature range can prove critical to real-world deployment.
- Pins and ports are a device interface's physical presences. From our past deployment experience, we've found that wiring is no trivial task. System integrators need explicit guidance when they connect sensors to sensor platforms, so unlike other standards, DDL made the choice to include pins and ports descriptions, despite the fact that they're of little relevance to services.
- Signals and protocols are the "raw readings" from a device interface. They can be an analog-to-digital converted (ADC) value from an analog output or a string from a digital port. Without defining the conversion or parsing method, these data carry no meanings. This layer is also a boundary at which the physical world ends and the digital world starts. Standards such as DeviceKit and IEEE 1451 choose this layer as the lower boundary of their scope.
- Measurements and commands are the semantics of signals and protocols. For example, when a temperature sensor sends out a signal (ADC value 500), it's then converted to measurement (115°F). These data are usually directly relevant to applications' interest, so this layer becomes the most important one in the description scope and is included by all sensor standards. SensorML and ECHONET assume that other standards have converted the signals and parsed the protocols, so they confine their modeling scope only to this layer.
- Networking protocol and configuration is the last item on the system integrator's checklist before the data are sent to middleware. However, whether this layer should be included in the standard is quite controversial because many sensors don't have networking capabilities. SensorML and DDL believe that network protocol processing is handled by sensor platforms rather than sensor themselves. Similarly, ECHONET specifies a protocol difference absorption block, which is part of the adapters rather than the devices.
- Applications and services are the sensor data's destination. Although there's no shortage of protocols standardizing interoperability between services, this layer hasn't been part of any sensor standard.

A quick scope analysis based on the aforementioned six layers reveals that measurement and control is the layer most common to all five standards.

2.5   DISSECTING DEVICES

Devices vary greatly in their physical shapes and forms as well as conceptually. People from various domains could hardly agree on a generic model that covers all devices, thus the standards we highlight here come from different research communities with different areas of concerns. It takes distinct perspectives to fully examine and analyze them.

2.5.1   Object-Oriented Perspective

One of the simplest ways to look at a device is to consider it an undividable object. Each device object has several internal properties and several methods to access them. Programming that follows the object-oriented paradigm is easy—an example is the ECHONET device specification. Each type of device is a class, with attributes and access rules specified in the document. Although object orientation is a straightforward idea, it is a powerful concept that provides many mechanisms.

2.5.2   Data-Oriented Perspective

In a data-oriented perspective, we can model a device as a process consisting of inputs, a data processing method, and output. Input and output data can be a phenomenon, a measurement, or a command—the device is simply a process that converts data from one form to another. For example, an electrical heater takes a command as its input, and creates a warmer environment (phenomenon)

TABLE 2.1:   The key differences among the five standards.

KEY COMPARISONS	ECHONET	IEEE 1451	SENSORML	DEVICE KIT	DEVICE DESCRIPTION LANGUAGE (DDL)
Encoding	Class specification in Plaintext	Interface Definition Language XML	XML	XML	XML
Design perspective	Object-oriented	Modular	Data-oriented	Modular	Data-oriented
Device model	Single object	Multiple blocks	Process chain	Multiple layers	Single device, cross layer

as the output. A collection of such linked processes forms a process chain that models a complex sensory system. SensorML is a typical example of this category. Originating from the remote-sensor community, which constantly works on data-intensive tasks such as geolocationing on a satellite image, SensorML specializes in modeling data processing. It supports customizable data types and lets users describe data processing methods with great efficiency.

TABLE 2.2: Comparison of five standards based on key criteria.				
OTHER COMPARISONS	**ECHONET**	**IEEE 1451**	**SENSORML**	**DEVICE KIT**	**DEVICE DESCRIPTION LANGUAGE (DDL)**
Basic component	Device	Block	Process	Device function layer	Device
Composite component	NA	Device	Process chains and sensor systems	Device	Derived virtual sensor (device)
Measurement modeling	Primitive data types	Complex data types	Complex data types	Primitive data types	Primitive data types and aggregate types
Protocol modeling	Inexplicit	Explicit	Inexplicit	Explicit	Explicit
Software support	NA	NA	NA	DKML language parser and Eclipse plug-in	DDL processor and Atlas bundle generator
Specification document	Published	Published	Published	Only schema available	Published online

2.5.3 Modular Perspective

In a modular perspective, we dissect a device into several modules, each of which is conceptually a self-contained component that provides certain functions. For example, in the IEEE 1451 standard, a device consists of three blocks (modules): a transducer, a function, and a Network Capable Application Processor (NCAP). A slight variation to the modular perspective is the layered design that DeviceKit uses. Three component layers divide a device's responsibility: a device layer provides the interface to hardware, a transport layer parses messages, and a connection layer supports I/O to the hardware. The modular design's benefit is that it provides a more organized and structured way to describe devices.

Also, similar devices can reuse modules easily. Table 2.1 summarizes the key differences among the five standards with respect to their encoding schemes, design perspectives, and device models. Table 2.2 shows additional comparisons based on other criteria.

Despite their early starts, neither SensorML nor IEEE 1451 has become the *de facto* industry/research standard. This is largely due to the absence of device manufacturers' involvement during the course of standardization. ECHONET, on the other hand, is endorsed by a small number of major manufacturers, but its impact has so far been confined to the Japanese domestic market. Therefore, although DeviceKit and DDL are much younger, they stand a good chance to promote sensor standardization to this community.

CHAPTER 3

Service-Oriented Device Architecture*

There is an increasing need to create interfaces between the physical world of sensors and actuators and the software world of enterprise systems. In this chapter, we cover a service-oriented device architecture (SODA) that represents a standing proposal for a set of specifications to enable device integration. Two sensor standards covered in Chapter 2 are based on SODA: the DeviceKit and the Device Description Language (DDL).

Let us first explain the specific choice of service orinetation in SODA. Wholesalers, retailers, and distributors demand immediate monitoring and control of shipments, enabled by RFID sensor data piped directly into their manufacturing, billing, and distribution software. Home healthcare monitoring can be implemented by providing devices such as EKG monitors and glucose monitors and pulse oximeters that can continuously monitor ambulatory patient status and alert healthcare providers of conditions requiring immediate care. A military control center must combine sensor data from various logistics and tactical environments—including the monitoring and control of RFID readers, vehicle control buses, GPS tracking systems, cargo climate controllers, and specialized devices—to provide situational awareness, preventive fleet maintenance, and real-time logistics.

The types of devices available for such scenarios continue to grow, while the cost of deploying them in the physical world and connecting them to all manner of networks continues to drop. However, the device interfaces, connections, and protocols are multiplying at a corresponding rate, and enterprise system developers are finding that integrating devices into the information technology (IT) world is daunting and expensive.

To eliminate much of the complexity and cost associated with integrating devices into highly distributed enterprise systems, a proposal led by IBM was put forth to leverage existing and emerging standards from both the embedded-device and IT domains within a Service-Oriented Device Architecture (SODA).

---

*This chapter is based on the following contribution: Scott de Deugd, Randy Carroll, Kevin E. Kelly, Bill Millett, and Jeffrey Ricker, "SODA: Service-Oriented Device Architecture," the Standards, Tools and Emerging Technologies Department, IEEE Pervasive Computing magazine, vol. 5, no. 3, 2006.

3.1   MODELING DEVICES AND SERVICES

SODA is an adaptation of a service-oriented architecture (SOA) [15] which integrates business systems through a set of services that can be reused and combined to address changing business priorities. Services are software components with well-defined interfaces, and they are independent of the programming language and the computing platforms on which they run [16–17]. The SODA approach to designing and building distributed software is to integrate a wide range of physical devices into distributed IT enterprise systems. At the simplest level, as Figure 3.1 shows, business services are used in today's enterprise SOAs.

SODA focuses on the boundary layer between the physical and digital realms. In other words, a sensor, such as a digital thermometer or an RFID tag reader, can translate a physical phenomenon into corresponding digital data. Or, an actuator, such as a heater relay or alarm beacon, can translate a digital signal into a physical phenomenon. Sensors and actuators combine either physically or conceptually to create complex devices and services—such as a climate-control module or a geo-fencing system that monitors a vehicle's position within a restricted area.

Viewing SODA in this way places few bounds on the device type or its usage, thus encompassing many types of devices from government, industry, and scientific enterprise. The devices range from basic sensor interfaces to complex diagnostic equipment. Whether providing the location of critical cargo, the blood-sugar level of a loved one, or the critical status of an energy distribution system, what devices share in common is that their data, functions, and events are critical services to the enterprise in which they're used [18].

We wouldn't need to think of devices in the broader terms of a digital realm without the types of complex distributed systems enabled by ubiquitous networks such as Bluetooth and the Internet. Without such networks, a signal would exist inside a single isolated machine or a well-defined isolated system. Networks enable a single processor to access signals from any number of devices. Thus,

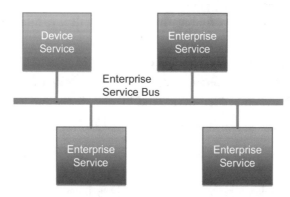

FIGURE 3.1: When modeled as services, device access and control can be made available to a wide range of enterprise application software using service-oriented architecture mechanisms.

the challenge of wide-scale device integration is predicated on the existence of a universal network capable of supporting complex distributed systems.

Before the networking capabilities enabled by the Internet, two or more devices would be associated only within a single well-defined system. The engineers would define the devices' interfaces and interdependencies in the system design. With the Internet, enterprise systems can now access signals from numerous devices on an ad hoc basis. The ability to access and control aspects of the physical realm, which is critical to an enterprise, opens new opportunities and advantages but can be self-limiting. The protocols, connections, and interfaces to devices are extremely diverse, and programming to them is often unfamiliar territory to system and Internet developers. Ancillary device interface software often is as critical to the completion of the IT project, with limited reuse and relatively high maintenance and support costs.

3.2   HOW A SERVICE-ORIENTED ARCHITECTURE CAN HELP

SODA aims to

- provide higher-level abstractions of the physical realm,
- insulate enterprise system developers from the ever-expanding number of standard device interfaces, and
- bridge the physical and digital realms with a known service or set of services.

SODA implementations can use existing and emerging device standards and SOA standards. SODA should be straightforward to implement, unlocking the potential to enable a new level of Internet-based enterprise systems.

Consider, for example, an enterprise scenario to process shipments of hazardous biological material. The application must track the materials during manufacturing and shipping, monitor and control their environment during shipment, and verify their delivery. Enterprise-system developers must build numerous complex distributed device interfaces for such an application:

- RFID readers and various digital I/O installed in the factory, delivery vehicles, shipping container, and at the receiving site;
- GPS devices on the shipping container and transport vehicle;
- a climate-control system in the shipping container;
- an interface to a vehicle control bus to determine the transport vehicle's status;
- various wired and wireless networks and protocols connecting all these devices, and numerous versions of all these items, depending on the specific device and on the protocol and networking vendor contracted by the system integrator.

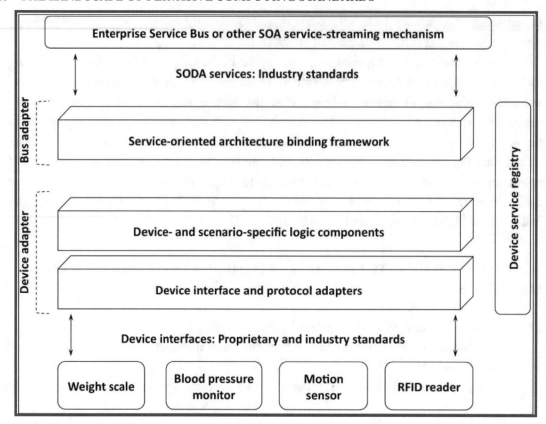

FIGURE 3.2: A Service-Oriented Device Architecture (SODA) implementation provides an abstract services model of a device by talking to proprietary and standard device interfaces on the one side and, on the other, presenting device data as SOA services over a network through a bus adapter.

In contrast, consider the enterprise system application developer's job if he or she could design an application to register interest in, and control, device services that can confirm, using an RFID-tagged item, that a shipment has cleared quality control, safety inspection, and the loading dock;

- confirm that the tagged shipment is secured in the shipping container in its assigned vehicle;
- provide a named vehicle and container's geographic location;
- provide notification of vehicle breakdown or of cargo leaving its geofence boundary;
- provide cargo's environmental status and listen for climate control adjustments;
- provide status of vehicle safety and mechanical systems; and
- confirm the environmental transit log and compare a shipping notice to a tagged shipment secured at a delivery location.

A device integration developer would be responsible for encapsulating devices as services, dealing with the device-specific connections and protocol as well as with network interfaces needed to publish the data over a defined SOA protocol. A standard specified device service can have a wide variety of underlying hardware, firmware, software, and networking implementations that don't affect the consumer of the service.

The overall system design would specify the required service interfaces. Suppliers would be responsible for the device adapters and service logic required to provide the specified service for their devices. Other developers could build more complex or composite services from lower-level device services. The system integrator could bid out components from multiple suppliers and avoid maintaining multiple versions of device-specific interfaces in the application code. Enterprise developers could code to a common or even standard set of services. They could not only build this application with clean device interfaces but also build and integrate future applications and system enhancements reusing these same device services. The enterprise could upgrade device hardware, firmware, and even the lower-level device interfaces with little or no impact on the consuming applications.

3.3   THE SODA ARCHITECTURE

SODA should be consistent with an enterprise SOA. An SOA provides the ability to map business processes and events across an open communication infrastructure, sharing data and functionality in an open, flexible manner [19]. We can apply SOAs' modular, componentized nature not only to traditional enterprise services but also to events and services from all data sources and devices, including embedded sensors and actuators and complex equipment.

Conventional approaches to device integration often center on custom interface software communicating to enterprise applications through a variety of IT middleware and API technologies. This approach has served enterprises well, but SOA, standards, and open software initiatives are moving beyond this middleware architecture. Although IT applications are being adapted to an SOA, standards for defining the low-level device interfaces are still emerging. However, technology exists today to leverage SOA across the entire spectrum of critical events and data originating from devices.

Mechanisms for building and sharing service interfaces, capabilities for remote software maintenance, and loosely coupled messaging models [20] present highly effective technologies for SODA's implementation. SODA requirements include:

- using a device adapter model to encapsulate device-specific programming interfaces;
- employing loosely coupled messaging on the services side, capable of supporting multiple streaming services commonly used in SOA enterprise systems, such as an Enterprise Service Bus;
- using open standards, where available, at the device- and services-interface level;

- providing a means to present standard or open service interfaces to devices that have proprietary protocols or where it might not be practical to drive standards into the low-level device interface;
- supporting the implementation of a spectrum of device adapters—from simple, low-cost sensor data to complex device protocols;
- supporting loading of remotely configurable logic components in device adapters for maintenance, upgrade, and extended functionality; and
- adapting security mechanisms as required for the domain.

A SODA implementation comprises three main components (see Figure 3.2). Device adapters talk to device interfaces, protocols, and connections on one side and present an abstract services model of the device on the other. The bus adapter moves device data over network protocols by mapping a device service's abstract model to the specific SOA binding mechanism used by the enterprise. The device service registry provides for the discovery and access of SODA services.

Where device interface standards don't exist, device interface and protocol adapters within SODA implementations provide a common model of devices to the software used to create service interfaces. Widespread adoption of standards at the device-interface level will reduce the development and maintenance costs of device adapters and their corresponding SODA services. However, standards at the services layer can provide the largest leverage for both the device and enterprise markets.

Rapid standardization of device services, device-services transport mechanisms, and tooling will let device manufacturers develop their interfaces and provide SODA services to the enterprise, shifting development responsibility for device adapters and services to the appropriate point in the supply chain rather than forcing enterprise developers to deal with thousands of APIs. For this adoption to take place, the SODA model must evolve with open and accessible standards, which must cover the specific services used within and across enterprises. Reducing the barriers to acceptance requires that the standards be open and part of a community and that samples, examples, frameworks, and tooling be made available through reference implementations. An open community of professionals can be developed by having government, corporate, and academic entities jointly sponsor such standards and activities that result in prototypes, pilots, and deployed solutions.

CHAPTER 4

Sensor Platforms*

Gordon Bell has predicted that, roughly every decade, a new class of computers or platforms is formed serving a new need (Bell's law [26]). The evolution from mainframes, to minicomputers, to connected workstations, to the PC clearly obeyed his law. Further to his credit, and driven by mobility needs, the laptop emerged in early 1990 and the PDA and smart phones have followed a decade later. And today, we witness the emergence of *sensor platforms*—a new class of tiny computers that fundamentally interconnects and bridges our cyber and physical worlds at a massive and pervasive scale.

By sensing the physical world, sensor platforms provide our cyber infrasturcture with critical data and unique opportunities to solve important, real-life problems. Additionally, sensor platforms empowers our cyber infrastructure to access, sense, and control the physical world wherever these platforms are embedded or deployed. Moreover, without any cyber infrasturctures, sensor platforms are by and among themselves capable of smartening up a physical space into what can be called a smart space or an intelligent environment.

We review the origins of the sensor platform and cover several key developments currently shaping this new breed of computers and its emerging applications.

4.1   THE BIRTH OF THE SENSOR PLATFORM

Mainstream research in sensor networks began around the mid-1990s with a number of DARPA-funded research initiatives undertaken by top US universities. The goal of this research was to design and create tiny autonomous computers, called sensor platforms, that could unobtrusively observe their environment through built-in sensors and report back to a remote base station. The primary use cases involved scattering hundreds or thousands of these sensor platforms in an area and tasking them with monitoring vehicular movement or environmental conditions and periodically reporting the data. To achieve such a goal, researchers envisioned a number of basic characteristics (requirements) that would set these sensor platforms apart from other computers available at that time.

---

*This chapter is based on the following contribution: Raja Bose, "Sensor Networks—Motes, Smart Spaces, and Beyond," the Standards, Tools and Emerging Technology Department, IEEE Pervasive Computing magazine, vol. 8, no. 3, July–September 2009.

4.1.1   Early Requirements of Sensor Platforms

- *Primitive processing capability*: since the sensor platforms weren't designed for general-purpose computing they required primitive yet robust processing capabilities. Unlike desktop PCs and servers, the platform would typically be designed around a low-cost microcontroller with limited processing and memory resources.
- *Self-organized networking*: the sensor platforms were designed for deployment in areas without existing network infrastructures so developing new networking capabilities became an important focus of sensor network research. Researchers envisioned that the platforms would collaborate to set up ad-hoc networks for routing sensor data and relaying it to a remote base station.
- *Low power operation*: the sensor platforms had to be self-sufficient and capable of operating on battery power for a reasonably long time. This implied that low-power consumption was more important than high throughput for both processing and networking hardware.
- *Tiny form factor*: the sensor platforms had to be small enough to remain unobserved and be easily deployed in large numbers without being conspicuous. This was publicized as the "smart dust" concept where sensor platforms could become so tiny that they'd be as ubiquitous and observable as mere specks of dust.

Due to the nature of research funding, the initial target applications favored relatively simple monitoring tasks where data were simply collected and stored, or aggregated in some form and transmitted out of the sensor network. Applications were written as part of the platforms' firmware with a focus on execution efficiency and resource management rather than programmability and ease of application development.

Mechanical actuation wasn't a primary application requirement, possibly because the targeted applications were more concerned with passive activities such as sensing and monitoring. Moreover, due to the low-power operation requirement, mechanical actuation was effectively out of the question for most use cases.

Let us take a look at the most popular first generation of sensor platforms.

4.1.2   Berkeley Motes

Berkeley Motes (Figure 4.1) are probably one of the best-known sensor platforms in the sensor network research community primarily due to their early commercialization by CrossBow Technologies [27].

Each mote was essentially made up of a sensor board with integrated sensors and a processing board consisting of a wireless radio transceiver and a microcontroller. All the motes were designed

FIGURE 4.1: Berkeley Mote with processing board, onboard sensors, and AA battery pack. The Mote is essentially a small form factor connected computer with self-contained processing, sensing, and power resources. Commercial versions of the Berkeley Motes include Rene, Mica, Mica2, Mica2Dot, MICAz, MICA2dot, Telos and iMote2.

to run off standard AA or lithium-ion batteries and in optimal conditions could last more than a year off a single battery charge (of course this depends on the used duty cycle). The processing component typically consisted of an Atmega128 RISC microcontroller from Atmel, though more recent versions (such as Telos and iMote2) used more powerful microcontrollers such as the MSP430 from TI and X-Scale from Marvel. The networking hardware used in different motes consisted of a number of low-power radios ranging from the Chipcon CC868 running proprietary ad-hoc networking stacks to standardized 802.15.4 radios such as the Chipcon CC2420, which supported ZigBee mesh networking stacks.

One of the motes research team's major contributions has been the TinyOS operating system, which was billed as the operating system for the sensor and sensor networks. TinyOS is an open source, component-based, embedded operating system developed primarily for the motes. It provides a set of software components that allows applications to interact with the processor, network transceiver, and the sensors. Applications written for TinyOS are compiled and statically linked with TinyOS code and loaded on the flash memory of the microcontroller for execution. The main goal of TinyOS is efficient application execution in resource constrained devices. It wasn't intended for easy and rapid application development or for supporting sophisticated applications simultaneously sharing the same set of sensing resources.

4.2   SENSOR PLATFORMS FOR SMART SPACES

Though sensor platforms were initially designed for deployment in primitive, infrasturcture-less environments, they are of immense utility in environments where networking and power resources are readily available, such as homes, offices and hospitals, to mention just a few. Thus, sensor platforms could integrate into existing spaces through their networking and power infrasturcture, and make them smart spaces. The smart home concept is probably the most enduring example of this idea.

Since research in sensor networks initially tended to focus on network connectivity and power consumption, application development remained a tedious affair. However, as the smart spaces concept evolved, it became clear that significant research was needed to tackle the problems of integration, programmability, and in general, ease of application development. Hence, a number of new desired characteristics for sensor platforms emerged:

- *Ease of integration.* The sensor platform has to be capable of integrating sensors and actuators into the smart space to make them easily accessible to application developers. This would enable software developers with little or no embedded systems experience to create applications using sensing and actuation resources available in the space. In existing platforms such as Berkeley Motes, the application and the underlying operating system layers are tightly coupled, making application development difficult for regular software developers.

- *Ease of programmability.* It should be possible to upgrade or modify smart space applications running on the sensor platforms unobtrusively and with minimal effort. This is an essential requirement for smart spaces since the requirements and capabilities of any real-world space will evolve over time. Existing sensor platforms such as Berkeley Motes require recompilation and reflashing of the entire firmware image for even minor modifications, hence they were inherently unsuitable for large-scale practical deployments. This requirement underlines an important need for application agnostic firmware with a strict separation between applications and operating system components.

- *Ease of application development.* In first-generation sensor platforms, applications were compiled as part of the firmware and executed entirely on the platform's microcontroller. This wasn't a big issue for earlier monitoring-based applications. However, in smart spaces, application development raises additional requirements in terms of programming methodologies and processing capabilities. Applications should be developed using standard models, tools, and IT expertise. This requires the sensor platform to provide clear, concise interfaces of its onboard sensors and actuators to the outside world. Such interfaces could be in the form of an API or a software proxy (e.g., a service or an object representation). Applications should also be allowed to execute mostly outside the sensor platdforms on powerful computers as needed.

- *Mechanical actuation.* The sensor platform has to support actuators such as switches, servos, motors, and invocation of other services that would allow it to control various aspects of the smart space.

- *Standardized communication protocols.* Unlike traditional sensor platforms that used proprietary or emerging ad-hoc networking protocols, the sensor platform needs to support standardized protocols such as IP to integrate seamlessly and effortlessly with the smart space's existing IT infrastructure.

We now present four sensor platforms that address the aforementioned additional requirements to varying extents.

4.2.1 Phidgets: Physical Widgets

Researchers at the University of Calgary developed an innovative sensor and actuation platform called Phidgets or "physical widgets" [28]. Their goal was to package physical sensors, actuators, and associated control software into physical widgets which would provide the same level of abstraction for physical devices that software widgets provide to an application developer. This would enable programmers to concentrate on developing end-user applications rather than getting distracted with low-level hardware and software issues.

The Phidgets hardware consists of an interface board with a microcontroller and USB connectivity that allows developers to hook up sensors and actuators (see Figure 4.2). A PC application can then access these sensing and actuation devices through USB. Phidgets provide intuitive software control libraries for different sensors and actuators in programming languages such as Visual

(a) Phidget platform connecting analog sensors to a host computer through a USB connection.

(b) Variety of analog sensors that connect to the Phidget Analog-USB platform.

FIGURE 4.2: Phidgets Platform with USB connectivity attached to a temperature sensor. Phidgets were one of the first sensor platforms to support mechanical actuators such as servos.

Basic where even novice programmers can simply drag and drop controls into their project and quickly create fairly sophisticated applications with a few lines of code. This is in sharp contrast to older platforms like the Berkeley Motes which required application developers to have embedded systems knowledge and write relatively low-level code which ran directly on the platform microcontroller, making it extremely hard to write and modify complex applications.

Phidgets are known for their ease of use and support of rapid prototyping, but they only support a USB communication interface and don't have any wireless or wired networking capability. This tethering makes it difficult to deploy sensors and actuators throughout the space without requiring several full-fledged PCs nearby. It makes it impractical to use the Phidgets hardware for large deployments since that would entail a messy arrangement of USB cables. Furthermore, Phidgets can't communicate among themselves (something that platforms such as Berkeley Motes support), and the application developer has to design any such interaction from scratch. Despite these shortcomings, they remain one of the most popular platforms for prototyping in the smart spaces research community.

4.2.2   The Atlas Sensor Platform

The Atlas sensor platform [11], developed at the Mobile and Pervasive Computing Laboratory at the University of Florida, was one of the first sensor platforms to utilize Service-Oriented Architecture (SOA) for automatic self-integration of sensors and actuators into smart spaces. Unlike other sensor platforms which either consisted solely of a set of distributed sensor nodes (such as Berkeley Motes) or a set of hardware adapters driven by PC-based software (such as Phidgets), Atlas is composed of semi-autonomous hardware sensor nodes (see Figure 4.3), coupled with an SOA-based backend middleware running on a PC or server machine.

The Atlas hardware is designed around the Atmel Atmega128 RISC microcontroller. It uses a modular architecture to separate sensors and actuators and their interconnections from the processing and communication layers. Unlike most other sensor platforms, it supports different communication and sensor and actuator interfaces using the same processing hardware and firmware.

Atlas was one of the first widely available sensor platforms that didn't provide support for ad-hoc networking and instead relied on the IP protocol with further upgrades providing ZigBee and USB capabilities. This enabled application developers to intergrate Atlas-based sensor networks into existing IT infrastructure. Any Atlas hardware node could be configured for attaching a number of sensors and actuators by using an intuitive Web-based interface. This is in sharp contrast to sensor platforms such as the Berkeley Motes which require firmware modifications to accommodate different sensor configurations.

Once an Atlas node is configured and powered on, all the sensors and actuators attached to it are abstracted into individual software services activated in a backend SOA framework such as OSGi. Making physical devices, sensors, and actuators available as software services enables appli-

FIGURE 4.3: Atlas Platform with WiFi communication board. Atlas Platform nodes feature application agnostic firmware with support for multiple swappable network interfaces such as WiFi, ZigBee, Wired Ethernet, and USB. Atlas also supports multiple swappable sensor/actuator boards.

cations residing in the smart space to dynamically discover and access them. Similar to the Phidgets concept, Atlas abstracts away low-level device details and provides high-level APIs to application developers for controlling sensors and actuators deployed in the space. Unlike Phidgets, Atlas packages its API's as services following the SOA archituecture. Also unlike Phidgets, Atlas supports a number of networking interfaces such as WiFi, ZigBee, Wired Ethernet, and USB, hence a heterogeneous network of Atlas nodes can be deployed over a smart space. Since Atlas abstracts away the networking details, all nodes appear as part of the same logical network even if they are on separate physical networks. Moreover, even though the Atlas platform's firmware is application agnostic, applications can push down basic data processing tasks onto the nodes such as push directives, aggregation, filtering, and event-based triggers involving multiple nodes across the network.

The utilization of SOA for abstracting physical devices into software services provided Atlas two major advantages. First, since each device is a high-level software service it enables developers to compose the same set of sensing and actuation devices into multiple applications without worrying about low-level embedded system details (integrate once, deploy everywhere). Second, the SOA paradigm in conjunction with the use of IP-based networking allows Atlas to easily integrate sensors and actuators into existing business process management and IT systems. This has always been the Achilles heel of most sensor platforms which meant that for the longest time, their utility was largely confined to prototyping, test-beds, and academic projects. Incorporation of these design features made Atlas extremely popular with commercial users who prefer standardized communications and interfaces for deploying sensors and actuators in their system.

4.2.3   Sun SPOT

The Small Programmable Object Technology (SPOT) project was lunched by Sun Micrososystem's research arm (Sun Labs) in 2005 [41] to create a demonstration for the Internet of Things. SPOTS are sensor platforms that allow programmers to develop Java applications for wireless sensor networks using standard development environments (e.g., NetBeans and Eclipse). The platform itself is Java-based, running a small J2ME visrtual machine called Squawk, which itself is written mostly in Java.

The SPOT sensor platform is shown in Figure 4.4. It consists of two stacked layers: the main processor/communication layer and the sensor I/O layer. The first layer hosts a powerful 32 bit ARM7 processor with 256K RAM and 2M Flash memory and can be powered using a battery or a USB connection. The same layer also hosts a Chipcon CC2420 802.15.4 radio with an integrated patch antenna. The sensor I/O board houses pre-selected sensors (3D acceloromter, light, and temperature) and offers 9 I/O pins for other sensor connections, limited to digital sensors only. The I/O board also provides a few high output current pins which is intended to allow SPOT to connect to actuators, and not just sensors.

A SPOT sensor network consists of several SPOT nodes, with one node acting as a gateway to other networks (other SPOT network, an Intranet or the Internet). A Gateway node must be connected to a computer through a USB interface.

Similar to the Berkeley mote, SPOT is concerned with programming the sensor platform itself. Unlike the mote, which focuses centrally on power issues, the SPOT focuses on the ease of programming as well as "up level-programmability" of the platform using exisiting powerful development tools. Power awareness of the SPOT platform is not ignored but was not the central

FIGURE 4.4: Sun SPOT sensor platform.

area of concern from the start. Compared to the Atlas platform, SPOT provides higher up-level programmability of the platform itself, but falls short in providing adequate infrastircture for automatic integration. SPOT also provides limited and less standardixed networking compared to the Atlas platform.

SPOT is nontheless a very interesting platform because it is a compltely open source project. The future of SPOT is not clear at the moment followingthe Oracle aquision of Sun Microsystems and its Sun Labs division.

4.2.4 Smart-Its

The Smart-Its project [42] was a collaborative effort between ETH Zurich, Lancaster University, the University of Karlsruhe Interactive Institute and the Valtion Teknillinen Tutkimuskeskus (VTT), funded by the European Union's Disappearing Computer initiative. Smart-Its (Figure 4.5) are specialized sensor platforms with hardware characteristics similar to the motes but packaged in a smaller form factor. The device typically has two versions, one consisting of an Atmega103L microcontroller from Atmel integrated with an Ericsson Bluetooth radio and the other featuring a PIC18F252 microcontroller integrated with a BiM2 radio transceiver using a proprietary networking stack.

Despite sharing similarities with Berkeley motes in hardware and basic characteristics, the Smart-Its sensor platform was created for a totally different purpose. On one hand, the motes evolved from the smart dust concept where researchers could deploy massive numbers of tiny sensors to observe certain phenomena and report back. On the other hand, Smart-Its were designed to

FIGURE 4.5: Smart-Its node with TR1001 wireless transceiver. The small form factor is intended to allow it to be easily attached to everyday objects as a tagging mechanism.

digitally tag and interconnect specific everyday objects (such as tea cups, keys, and toys)—they are not designed as general purpose sensor platforms. They act as smart tags that allow dumb objects to sense and interact with each other digitally.

One of the interesting uses of the Smart-Its platform has been for what its inventors called context-proximity-based connection of artifacts. In this usage, objects tagged with Smart-Its platforms automatically pair up with other tagged objects in their proximity when both exhibit similar sensory context. For example, if two Smart-Its tagged objects are shaken while they are within radio range of each other, they automatically pair up and can interact with each other. This is essentially done by sampling different sensors on the Smart-Its platform (for example, the accelerometer) and broadcasting those values over a common communication channel so that other Smart-Its in the vicinity can check if they also satisfy the same context and then pair up with the sender.

CHAPTER 5

Service Discovery and Delivery Standards*

For the past ten years, competing industries and standards developers have been hotly pursuing automatic configuration, now coined the broader term service discovery. Jini, Universal Plug and Play (UPnP), Salutation, and Service Location Protocol have been among the front-runners in this new race. However, service discovery goes beyond the need for plug-and-play solutions and support for the small office/home office (SOHO) user. Service discovery is very important in mobile and pervasive environments given the highly dynamic nature of the devices and the users. Also, many of the emerging sensors and device standards (Chapter 2) and the SODA device archicture framework driving these stadards (Chapter 3) are service-oriented, which benefit significantly from service discovery mechanisms and standards.

The case for mobile computing. Mobility means getting away from configured environments and into foreign networks with unknown infrastructures. However, because a mobile computer can't predict such infrastructures, it might not know to take advantage of them or even have the capabilities to interact with them. For example, a mobile computer might not be able to use a nearby printer because it does not have the appropriate printer driver, or perhaps a PDA will experience slow Web access because it is not aware of a nearby Web proxy caching server. As mobile computing evolves beyond the ability to wirelessly connect to read email or surf the Web anywhere and on any device, it is bound to exploit local resources, peers, and services. With the advent of location-based services and peer-to-peer computing, service discovery is taking on a new role as a critical middleware for mobile computing, and is enabling "opportunistic" new models of programming and application development.

The case for pervasive computing. Service discovery also benefits pervasive computing environments, where numerous computing elements, sensors, actuators, and users often must interact

*This chapter is based on the following contribution: Sumi Helal, "Standards for Service Discovery and Delivery," the Standards, Tools and Emerging Technologies Department, IEEE Pervasive Computing magazine, vol. 1, no. 3, July–September 2002.

to achieve the desired functionality and intelligence. In such environments, self-advertisement and peer discovery can enable the pervasive space to autonomically integrate, and to dynamically change and evolve without major system reengineering or reconfiguration. This is especially suitable as mentioned earlier under the SODA device architecture.

5.1   SERVICE DISCOVERY PROTOCOLS

Consider the following three scenarios. First, imagine finding yourself in a taxi without your wallet. Fortunately, you have a Jini technology-enabled cellular phone, and your cellular provider uses Jini technology to deliver network-based services tailored to your community. On your phone screen, you see a service for the City Cab Company, so you download the electronic payment application to authorize paying your taxi fare. The company's payment system instantly recognizes the transaction and sends a receipt to the printer in the taxi. You take the receipt, and you're on your way.

Second, consider an insurance salesman who visits a client's office. He wants to brief the client on new products and their options, which are stored in his Windows Mobile handheld PC. Because his handheld PC has Wi-Fi and supports UPnP, it automatically discovers and uses an Ethernet-connected printer without any network configuration and setup. He can print whatever he wants from his handheld and promote the new products.

Finally, consider an intelligent, online overhead projector with a library client. After being authenticated, the user might select a set of electronically stored charts or other documents for viewing. Rather than bringing transparencies to a meeting, the user accesses them through the LAN server in the library.

Scenario 1 is a Jini demo scenario from Sun Microsystems, Scenario 2 is a UPnP scenario from Microsoft, and Scenario 3 comes from Salutation. At a glance, they all seem to talk about the same stories: mobile devices, zero configuration, impromptu community enabled by service discovery protocols (SDPs), and cooperation of the proximity network. Without mentioning the trademarks, we would hardly know which company is telling which scenario. These SDPs, however, have different origins. They see the problem from different angles and have different approaches for solving it.

We will next cover Jini, Universal Plug and Play (UPnP), Service Location Protocol, and Bluetooth SDP in some details. Salutation which somehow became extinct a few years ago will not be covered in this chapter.

5.2   JINI

Sun Microsystems introduced Jini, based on the Java technology, in 1998. The heart of Jini is a trio of protocols: discovery, join, and lookup. A pair of these protocols—discovery and join—occurs when you plug a Jini device into a network; discovery occurs when a service looks for a lookup service with which it can register, and join occurs when a service locates a lookup service and wants to join it. Lookup occurs when a client or user locates and invokes a service described by its interface type (written in the Java programming language) and possibly other attributes. For a client in a Jini community to use a service:

- The service provider must locate a lookup service by multicasting a request on the local network or a remote lookup service known to it a priori (see Figure 5.1a).
- The service provider must register a service object and its service attributes with the lookup service. This service object contains the Java programming language interface for the service, including the methods that users and applications will invoke to execute the service, along with any other descriptive attributes (see Figure 5.1a).
- A client then requests a service by Java type and perhaps other service attributes. The lookup server ships a copy of the service object over the network to the client, who uses it to talk to the service (see Figure 5.1b).
- The client interacts directly with the service via the service object (see Figure 5.1b).

Jini technology consists of an infrastructure and a programming model that address how devices connect with each other to form an impromptu community. Jini uses the Java remote method invocation (RMI) protocol to move code around the network.

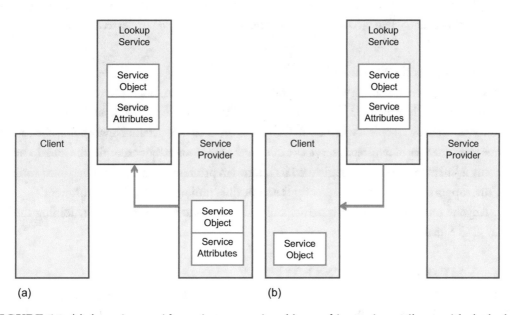

(a)   (b)

FIGURE 5.1: (a) A service provider registers a service object and its service attributes with the lookup service. (b) A client requests a service from service attributes, and a copy of the service object moves to the client.

We can view the Jini lookup service as a directory service or broker. Jini uses three related discovery protocols. When an application or service first becomes active, the multicast request protocol finds lookup services in the vicinity. Lookup services use the multicast announcement protocol to announce their presence to services in the community that might be interested. The unicast discovery protocol then establishes communications with a specific lookup service known a priori over a wide-area network.

However, a Jini lookup service is not just a simple name server. It maps the interfaces that clients see to service proxy objects. It also maintains service attributes and processes match queries. Clients download the service proxy, which is usually an RMI stub that can communicate back with the server. This proxy object lets clients use the service without knowing anything about it. Hence, there is no need for device drivers in case the service provided is a device (such as a printer). Although service proxy objects represent a typical scenario of service invocation, the downloaded service object can be the service itself or a smart object capable of speaking in any private communication protocol.

5.2.1   Leasing in Jini

Jini grants access to its services on a lease basis. A client can request a service for a desired time period, and Jini will grant a negotiated lease for that period. This lease must be renewed before its expiration; otherwise, Jini will release the resources associated with the service. Leasing lets Jini be robust and maintenance-free when faced with abrupt failures or the removal of devices and services.

5.2.2   Distributed Programming in Jini

Besides the basic service discovery and join-and-lookup mechanism, Jini supports remote events and transactions that help programmers write distributed programs in a reliable and scalable fashion. Remote events notify an object when desired changes occur in the system. Newly published services or some state changes in registered services can trigger these events. For example, the lookup service can notify a Jini palmtop that has registered its interest in printers when a printer becomes available. Also, Jini supports the notion of transactions and flexible notions of atomic commitment.

Anyone interested in Jini can participate and contribute to the standard by joining the Jini Forum [21]. Sun Microsystems acts as the steward for this forum.

5.3   UNIVERSAL PLUG AND PLAY

UPnP is an evolving Microsoft-initiated standard that extends the Microsoft Plug-and-Play peripheral model. It aims to enable the advertisement, discovery, and control of networked devices, services, and consumer electronics. In UPnP, a device can dynamically join a network, obtain an IP address, convey its capabilities on request, and learn about the presence and capabilities of other devices. A device can also leave a network smoothly and automatically without leaving any unwanted state behind. UPnP leverages TCP/IP and Web technologies, including IP, TCP, UDP, HTTP, and XML. It uses the protocol stack in Figure 5.2 for service discovery, advertisement, description, and eventing.

5.3.1   Joining and Discovery in UPnP

UPnP uses simple service discovery protocol (SSDP) for service discovery. This protocol announces a device's presence to others and discovers other devices or services. Therefore, SSDP is analogous to the trio of protocols in Jini. SSDP uses HTTP over multicast and unicast UDP, referred to as HTTPMU and HTTPU, respectively.

A joining device sends out an advertisement (ssdp:alive) multicast message to advertise its services to control points. Control points function similar to Jini's lookup services. A control point, if present, can record the advertisement, or other devices might also directly see this multicast message. In contrast to Jini, UPnP can work with or without the control points (lookup service). It sends a search (ssdp:discover) multicast message when a new control point is added to a network. Any device that hears this multicast will respond with a unicast response message.

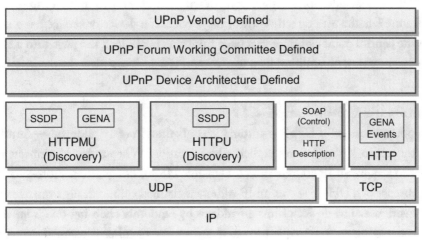

GENA: General Event Notification Architecture
SOAP: Simple Object Access Protocol
SSDP: Simple Device Discovery Protocol
UDPs: User Datagram Packets
UPnP: Universal Plug and Play

FIGURE 5.2: Universal Plug and Play protocol stack.

UPnP uses XML to describe device features and capabilities. The aforementioned advertisement message contains a URL that points to an XML file in the network that describes the UPnP device's capability. By retrieving this XML file, other devices can inspect the advertised device's features and decide whether it is important or relevant to them. XML allows complex and powerful description of device and service capability as opposed to Jini's simple service attribute.

5.3.2   UPnP Service Description

After a control point has discovered a device, it learns more about how to use it, control it, and co-ordinate with it by retrieving its XML description file. Control is expressed as a collection of Simple Object Access Protocol (SOAP) objects and their URLs in the XML file. To use a specific control, a SOAP message is sent to the SOAP control object at the specified URL. The device or the service returns action-specific values.

A UPnP description for a service includes a list of actions to which the service responds and a list of variables that model the service's state at runtime. The service publishes updates when these variables change, and a control point can subscribe to receive this information. Updates are published by sending event messages that contain the names and values of one or more state variables. These messages are also expressed in XML and formatted using the General Event Notification Architecture.

UPnP features an additional higher-level description of services in the form of a user interface. This feature lets the user directly control the service. If a device or service has a presentation URL, then the control point can retrieve a page from this URL, load the page into a browser, and (depending on the page's capabilities) let a user control the device or view the device's status.

5.3.3   Automatic Configuration of IP

Another important feature of UPnP is automatic configuration of IP addresses—AutoIP. It lets a device join the network without any explicit administration. When a device connects to the network, it tries to acquire an IP address from a Dynamic Host Configuration Protocol server. However, in the absence of a DHCP server, an IP address is claimed automatically from a reserved range for local network use. The device claims an address by randomly choosing one from the reserved range and then making an ARP request to see if anyone else has already claimed that address.

Headed by Microsoft, the UPnP Forum [22] oversees the standard's developments. The standard development process is similar to the Java Community Process. A significant number of OEM and consumer electronic products implement UPnP and are UPnP compliant. Figure 5.3 shows an example of an UPnP embedded web server by Zilog.

FIGURE 5.3: A Zilog's eZ80 Micro Web server running Metrolink IPWorks with UPnP support.

5.4   SERVICE LOCATION PROTOCOL

Service Location Protocol (SLP) is an Internet Engineering Task Force (IETF) standard for decentralized, lightweight, and extensible service discovery. It uses service URLs, which defines the service type and address for a particular service. For example, "service:printer:lpr://hostname" is the service URL for a line printer service available at hostname. Based on the service URL, users (or applications) can browse available services in their domain and select and use the one they want.

There are three agents in SLP: the *user*, *service*, and *directory*. The UA is a software entity that sends service discovery requests on a user application's behalf. The SA broadcasts advertisements on behalf of a service. As a centralized service information repository, the DA caches advertisements from SAs and processes discovery queries from UAs. An SA advertises itself by registering with a DA. The registration message contains the URL for the advertised service and for the service's lifetime, and a set of descriptive attributes for the service. The SA periodically renews its registration with the DA, which caches the registration and sends an acknowledge message to the SA. A UA sends a service request message to the DA to request the service's location. The DA responds with a service reply message that includes the URLs of all services matched against the UA request. Now, the UA can access one of the services pointed to by the returned URL. In SLP, the DA is optional. A DA might not exist in a small network, in which case the UAs' service request messages are directly sent to the SAs.

SLP supports service browsing and string-based querying for service attributes, which let UAs select the most appropriate services from among available services in the network. SLP lets

UAs issue query operators such as AND, OR, comparators, and substring matching. This is more powerful than Jini and UPnP service attribute matching, which can be done only against equality.

The SLP standard is accessible from the IETF SvrLoc working group Web site [23].

5.5   BLUETOOTH SDP

Unlike Jini, UPnP or SLP, the Bluetooth SDP is specific only to Bluetooth devices [24]. It primarily addresses the service discovery problem. It doesn't provide access to services, brokering of services, service advertisements, or service registration, and there's no event notification when services become unavailable. SDP supports search by service class, search by service attributes, and service browsing. The latter is used when a Bluetooth client has no prior knowledge of the services available in the client's vicinity. SDP is structured as a Bluetooth profile and runs on a predefined connection-oriented channel of the L2CAP Logical Link layer.

5.6   SUCCESSES AND FAILURES (2000–2010)

Jini. Not many Jini products are available on the market today. In 1999, a year after Jini's introduction, companies such as Epson, Canon, Seagate, and Quantum agreed to embed Jini in some of their product lines. However, later in the same year, these companies warned that it might take up to two years to accomplish this. People then predicted that Jini storage devices would be first to hit the market, but this never materialized. A short-term success with Jini was PsiNaptic, a medium size Canadian company specializing in pervasive computing middleware. The company offered Jmatos—a Jini product that supports Dallas Semiconductor's (now Maxim's) Tiny Internet Interface (known as Tini), which has an embedded Java Virtual Machine. Tini is a small-footprint microcontroller with a rich set of on-board interfaces. Unfortunately, Jmatos like many other middleware type of products struggled to position itself in a new shaping market.

UPnP. Unlike Jini, UPnP took a different turn. Many UPnP products are available in the market today. In addition to the tens of millions of Windows PCs (which come with UPnP support), hundreds if not thousands of OEM and consumer products support the UPnP protocol. Over 800 companies are members of the UPnP forum, each offering one or more UPnP products. Such products include storage devices, home audio/video, digital cameras, and video camcorders, set top boxes, home routers, smart phones, and many other business and home office equipments. In addition to products, many UPnP development kits (SDK) are available today that would let developers of devices, consumer electronics, and embedded systems build UPnP support into their products (e.g., Allegro Software, Virata, Intel, Lantronix, Atinav, Metrolink, Microsoft, and Siemens). Most of these SDKs support multiple platforms (Windows, Linux, Mac OS, etc.), and support multiple languages (C, C++, and Java).

Salutation. Even though now completely extinct, several Salutation Consortium products actually made it to the market. These products weremostly office automation products (fax machines, printers, copiers, and scanners). Another product was IBM NuOffice—a networked office system based on Lotus Notes, that lets users import and export data from and to any Salutation device.

SLP. Backed by IETF and aligned with other established protocols (including Lightweight Directory Access Protocol (LDAP), Domain Naming System (DNS), and DHCP), developers have widely accepted SLP as a simple, minimum requirement service discovery protocol. Another source of this acceptance is SLP's scope, because it attempts only to locate—not access or deliver—the service. SLP is used by Hewlett Packard's JetSend technology, which supports HP's office equipment and consumer electronics. Other vendors with SLP printer and network products include Axis, Lexmark, Xerox, Minolta, IBM, Novel, and Zephyr, and Axis also offers SLP storage devices. In addition to office and networking equipment, several platforms support SLP, including Sun, Caldera, Novel, and Apple.

Service discovery overall. Service discovery has come a long way to becoming a major standardization and development effort, but except for UPnP, the picture is not as impressive when we consider market acceptability and available products. In addition, in their current form and shape, most service discovery standards do not address all mobility's needs and the special requirements of pervasive spaces. Their full potential in mobile and pervasive environments is therefore yet to be unleashed.

Current SDPs are designed for use in local area networks. The IP multicast range, for example, limits discovery in Jini. This is inadequate for mobile clients requiring access to services from wide area networks.

Another problem with existing SDPs is their lack of support for mobile devices. For instance, Jini requires JVM and RMI capabilities on the client slide, which has hindered its widespread use on mobile devices. A quick fix to this problem was the Jini Surrogate Architecture introduced by Sun Micrososystems. Using surrogates, a device does not have to have or understand JVM or RMI. It only must be able to store and "squirt" Jini code that uses RMI to a proxy (the surrogate) on the local network to act on its behalf. Unfortunately, surrogates are more a solution to stationed devices than to mobile devices.

Another limitation of current service discovery frameworks is that they do not consider important context information. For example, there is no support to service routing and selection based on the client's location. Other unexploited contextual information includes distance to service, time, service load, and quality of service instances.

CHAPTER 6

The Open Services Gateway
Initiative (OSGi)*

Early pervasive spaces have been developed primarily with proprietary technology that lacked a long-term vision of evolution and interoperation. However, to culminate into a healthy industry, future pervasive spaces will inevitably be comprised of a wide variety of devices and services from different manufacturers and developers. We must therefore achieve platform and vendor independence as well as architecture openness if we hope for pervasive spaces to become common places.

The Open Services Gateway Initiative (OSGi) attempts to meet these requirements by providing a managed, extensible framework to connect various devices in a local network such as in a home, office, or an automobile. By defining a standard execution environment and service interfaces, OSGi promotes the dynamic discovery and collaboration of devices and services from different sources. Moreover, the framework is designed to ensure smooth space evolution over time and to support connectivity to the outside world, allowing remote control, diagnosis, and management.

As we examine the details of the OSGi standard, we will see how it nicely caters to the emerging requirements of pervasive spaces and sensor platforms. Specifically, it provides the needed support for programmability, self-integrration, ease of application development, and goes one step further to supporting the greater life cycle of the pervasive space.

Originated in 1999 to deliver WAN services to home environments, OSGi today offers a unique opportunity to pervasive computing as a potential framework for achieving interoperability and extensibility. Through two major enhancements over the last few years, the specification now includes numerous new services and features that support various usage models.

6.1   THE FRAMEWORK

The OSGi framework provides general purpose, secure support for deploying extensible and downloadable Java-based service applications known as bundles. An OSGi service platform is an

*This chapter is based on the following contribution: Choonhwa Lee, David Nordstedt and Sumi Helal, "Enabling Smart Spaces with OSGi," the Standards and Emerging Technologies Department, IEEE Pervasive Computing magazine, vol. 2, no. 3, July–September 2003.

instantiation of a Java virtual machine, an OSGi framework, and a set of bundles (see http://www
.osgi.org/Links/HomePage).

Running on top of a Java virtual machine, the framework provides a shared execution environment that installs, updates, and uninstalls bundles without needing to restart the system. Bundles can collaborate by providing other bundles with application components called services. An installed bundle might register zero or more services with the framework's service registry. This registration advertises the services and makes them discoverable through the registry so that other bundles can use them. The framework also manages dependencies among bundles and services to facilitate coordination among them.

The framework provides application developers with a consistent programming model by defining only interfaces between the framework and the services. The real implementation is left to the bundle developers; service selection and interaction decisions are made dynamically at service discovery and use time. This separation of service definition and implementation ensures that services from different service providers interoperate on the managed framework. Figure 6.1 depicts the OSGi service gateway placed between a WAN and a LAN.

We can deploy a bundle in an OSGi service platform to provide application functions to other bundles or users. The bundle is a Java Archive (JAR) file that contains Java class files to implement zero or more services, a manifest file, and resources such as HTML pages, help files, icons, and so forth. The manifest file describes the JAR file itself and provides additional information about the bundle. Some examples of manifest headers include *Import-Package*, *Export-Package*, and *Bundle-Activator*. Import-Package and Export-Package guide the framework to resolve shared package dependencies by listing package names that must be imported and those that can be exported. Packages are an indivisible unit constituting an OSGi bundle. The Bundle-Activator header specifies a class whose start and stop methods are called by the framework to start and stop the bundle.

A bundle can register services with the framework service registry. In this case, the service implementation (that is, the service object), which is represented by its Java service interface, is what actually gets registered. Bundles can discover services offered by each other by querying the service registry using a simple service discovery interface. When a bundle queries the registry, it obtains references to actual service objects registered under the desired service interface name. Besides the interface name, we can further describe services using a collection of key-value pairs. We can then match the services' properties against a filter parameter to narrow down the query results. The OSGi filter is based on the LDAP search filter's string representation.

The framework manages dependency among bundles that offer and use a given service. For example, when a bundle is stopped, the framework automatically unregisters all services that the bundle registered. Also, service events can notify a bundle when a service from other bundles is

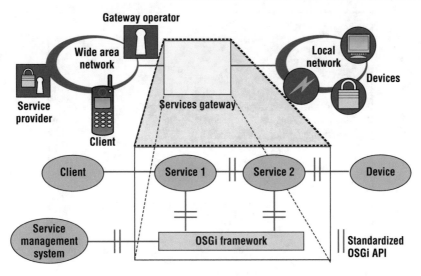

FIGURE 6.1: The OSGi framework and services.

registered, modified, or unregistered. Figure 6.2 shows how the service registry is used to advertise and discover services [25].

The simplified code fragments in Figure 6.3 are an implementation of a sample *Hello World* service along with the bundles that provide and use it. Figures 6.3(a) and 6.3(b) show the service interface definition and the service implementation. The bundle in Figure 6.3(c) registers the service implementation object with the framework service registry along with a service property. The bundle in Figure 6.3(d) shows how to discover the service object using the service interface name

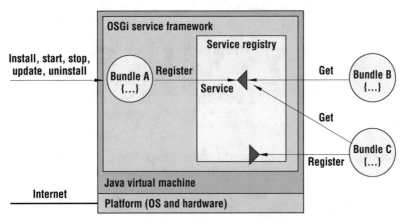

FIGURE 6.2: Bundle collaboration through service registry.

```
package srvs;
public interface HelloWorldService {
    public void foo();
}
```
(a) Interface

```
public HelloWorldServiceImpl implement HelloWorldService {
    public void foo() {
        System.out.println("Hello World");
    }
}
```
(b) Implementation

```
public class HelloWorldProvider implements BundleActivator {
    private ServiceRegistration reg = null;

    public void start(BundleContext ctxt) throws Exception {
        HelloWorldService fooSrv = new HelloWorldServiceImpl();
        Properties props = new Properties();
        Props.put("description", "sample");
        reg = ctxt.registerService("srvs.HelloWorldService", fooSrv, props);
    }

    public void stop(BundleContext ctxt) throws Exception {
        if (reg != null)   reg.unregister();
    }
}
```
(c) Provider Bundle

```
public class HelloWorldServiceUser implements BundleActivator {
    public void start(BundleContext ctxt) throws Exception {
        ServiceReference[] ref = ctxt.getServiceReference
            ("srvs.HelloWorldService", "(description=sample)");
        HelloWorldService fooSrv = (HelloWorldService)
        ctxt.getService(ref[0]);
        fooSrv.foo();
    }
    public void stop(BundleContext ctxt) throws Exception {
    }
}
```
(d) User Bundle

FIGURE 6.3: A sample *Hello World* OSGi service.

and service property. As shown, every bundle should implement the BundleActivator interface, which defines the start and stop methods.

6.2 SPECIFICATIONS

Since the OSGi specifications' first release in May 2000, they have gone through three major up-dates, in October 2001, March 2003, and September 2009. The latest specs (known as Release

4, version 2) provide several reference architectures, including a large service delivery network managed by an operator and a home or office network to integrate various computing devices. Although OSGi can apply to various scenarios, perhaps the most appealing use model is the remote-management reference architecture. It lets an operator manage a large network of service platforms and services supplied by different service providers. Remote managers are at a central site, and they communicate with manager agents on remote target service platforms. A manager agent is a set of bundles that provides remote management of the service platform to its central remote manager. Table 6.1 summarizes the services and classes that the latest OSGi specification (release 4.2) defines, all of which are available to bundle developers.

6.3 DEVELOPMENT TOOLKITS
The OSGi Web site (www.osgi.org) lists OSGi service platform developer kits available from several vendors. We look at a few from Prosyst, IBM, and Sun Microsystems.

6.3.1 ProSyst mBedded Builder
Prosyst has a complete line of products for an OSGi 2.0 system, which includes their mBedded Server, mBedded Remote Manager, and mBedded Builder. The mBedded Server runs on many operating systems and provides the OSGi service platform. The mBedded Remote Manager provides for easy management of a possibly large network of OSGi service platforms. The mBedded Builder is a complete, integrated development environment with many extensions for application developers to build services for an OSGi system. In addition to a complete GUI IDE for developing Java applications, there is support for CVS (Concurrent Versions System), custom device GUI design, and creating OSGi service bundles.

6.3.2 Java Embedded Server
Sun Microsystems produced its Java Embedded Server before the OSGi consortium was founded. In fact, Sun was a founding OSGi member, and JES significantly influenced the OSGi specification. JES has evolved to the current 2.0 version, which is fully compliant with OSGi 1.0. The current version is a free download at http://java.sun.com/jes and is supported on the Solaris 2.7 and Windows NT platforms. There is a free plug-in to the SunONE Studio IDE, also available on Sun's Web site.

6.3.3 IBM SMF Bundle Developer
IBM's Websphere Device Developer helps create mobile applications for many devices, including the J2ME applications that have become so popular in the past couple of years. This developer product now also supports IBM's own OSGi implementation, Service Management Framework, through the SMF Bundle Developer plug-in. This plug-in allows GUI creation and manipulation of OSGi bundles, manifest headers, and testing. The IBM SMF implements OSGi 3.0 but omits

many features, such as Start Level, URL Handlers Service, I/O Connector Service, Wire Admin Service, Namespace, Jini Driver Service, UPnP Device Service, and Initial Provisioning support.

6.4   ADOPTION AND PRODUCTS

Numerous member companies developing and supplying OSGi-based products back the OSGi consortium. The E2-Home Project in Stockholm executed the first commercial deployment of a consumer-oriented OSGi service platform using the Gatespace e-Services embedded software of Makewave (formerly Gatespace Telematics, http://www.makewave.com). This project, which launched in 2001, is used in 180 condominiums for services such as energy management, home automation, and community services. Members of the housing project can access the Web to monitor their utility consumption, reserve common areas such as the laundry and sauna, and order store items for delivery. All these services, including burglar alarm and email access, are available from a convenient touchscreen computer in the kitchen.

The makewave is a primary maintainer of Knopflerfish (KF) which is a universal open source OSGi Service Platform. Knopflerfish is developed based on Gatespace's OSGi framework GDSP source code. Currently makewave released Knopflerfish 3 beta-4 (KF3), which fully implements the core functions according to v4.2 specifications except for the declarative services compendium service. The Gator Tech Smart House (GTSH) of University of Florida uses the Knopflerfish to implement the Atlas middleware which enables programmable pervasive space [9].

Equinox is an implementation of the OSGi R4 specification, a set of bundles that implement various optional OSGi services and other infrastructure for running OSGi-based systems. Equinox is a module of Eclipse which is also developed by IBM. Eclipse is a popular open development platform comprised of extensible frameworks, tools and runtimes for building, deploying and managing software across the lifecycle (http://www.eclipse.org). Including Equinox is a step towards making Eclipse a runtime platform, and not just a development platform.

Cisco has deployed OSGi service platform technology in their CiscoWorks 2000 Service Management Solution, which manages service agreement levels between enterprise and other networks to ensure quality of service. Cisco is also using the Gatespace software in their system.

Whirlpool (www.whirlpoolcorp.com) uses IBM's OSGi service platform for its Home Solutions product line. Featured services will include Internet-enabled appliance controls, Internet connection sharing, and home firewall capabilities. WebPad inside the house or a WAP-capable cell phone from outside.

InterComponentWare uses the Prosyst OSGi technology in their LifeSensor product (www.lifesensor.com), which transfers medical information from patients' medical devices to remote caregivers.

TABLE 6.1: OSGi specifications.

SPECIFICATION (FIRST RELEASE)	DESCRIPTION
Log Service (1)	Provides two services—Log Service to record log information and Log Reader Service to retrieve the information.
HTTP Service (1)	Provides APIs for bundles to register servlets or resources such as static HTML pages, images, sounds, and so on, so that a standard Web browser can access them in an OSGi service platform.
Device Access (1)	Specifies a device model based on the device manager, which controls automatic attachment of a suitable device driver service to a newly registered device service.
Configuration Admin Service (2)	Configuration data is set of properties that a remote agent or other applications in an OSGi environment Admin maintain. A Configuration Admin Service instance hands over the configuration data to bundles on their registration or when their configuration changes at a later time.
Metatype Service (2)	Specifies interfaces that allow bundle developers to describe the type information of data. The data are based on attributes, which are key/value pairs like properties.
Preference Service (2)	Provides a bundle with persistent storage of named data values. Unlike the java.util.Properties class, it supports a hierarchical naming model, and its key/value pairs can be stored in a remote machine.
User Admin Service (2)	Defines a flexible authentication to adopt different authentication schemes. Once authenticated, a bundle uses its role-based authorization to verify if the user is authorized to perform the requested action.

(continued)

TABLE 6.1: (*continued*)

SPECIFICATION (FIRST RELEASE)	DESCRIPTION
Wire Admin Service (3)	Intended for user interfaces or management applications. Wire Admin Service controls the wiring of services—for example, wiring data-producing services to data-consuming services. It enables dynamically configurable collaboration among bundles.
I/O Connector Service (3)	Defines a flexible, extendable communication API based on the J2ME Connector framework of the javax.microedition.io package.
Initial Provisioning (4)	To allow freedom regarding the choice of management protocol, it defines how the Management Agent can make its way to the Service Platform, and gives a structured view of the problems and their corresponding resolution methods.
UPnP Device Service (3)	Device understands UPnP protocol to transform UPnP services to OSGi services and vice versa.
Declarative Services (4)	A procedural service model which enables the building of applications out of bundles that communicate and collaborate using these services. Several complications arise when the OSGi service model is used for larger systems and wider deployments. It addresses some of the complications such as Startup Time, Memory Footprint and Complexity.
Event Admin Service (4)	Allows to deal with events, either as an event publisher or as an event handler. It provides an inter-bundle communication mechanism. It is based on a event publish and subscribe model, popular in many message based systems.

Deployment Admin (4)	Provides mechanisms to manage the lifecycle of bundles, configuration objects, and permission objects. It standardizes the access to some of the responsibilities of the life-cycle management of interlinked resources on an OSGi Service Platform.
Auto Configuration (4)	Allows the configuration of bundles and defines the format and processing rules of an Autoconf Resource Processor. An Auto Configuration Resource contains information to define Configuration objects for the Configuration Admin Service Specification.
Application Admin (4)	It simplifies the management of an environment with many different types of applications which are simultaneously available. It enables applications that manage other applications, regardless of application type. This specification supports enumerating, launching, stopping and locking applications.
Device Management Tree (DMT) Admin Service (4)	DMT Admin Service defines an API for managing a device using concepts from the Open Mobile Alliance (OMA) DM protocol specifications. This API has been designed to be useful with or without an OSGi service platform.
Monitor Admin Service (4)	Allows management systems to receive information from applications and services to monitor the status of the device. It outlines how a bundle can publish Status Variables and how administrative bundles can discover Status Variables as well as read and reset their values.
Foreign Application Access (4)	Enables foreign applications that are not natively supported by foreign application models like MIDP,Xlets, Applets, other Java application models to participate in the OSGi service oriented architecture.

(continued)

TABLE 6.1: (*continued*)

SPECIFICATION (FIRST RELEASE)	DESCRIPTION
Blueprint Container (4)	Addresses decoupling issues to reuse modules. It defines a dependency injection framework, specifically for OSGi bundles, that understands the unique dynamic nature of services. It provides an OSGi bundle programming model with minimal implementation dependencies and virtually no accidental complexity in the Java code.
Tracker (2)	A utility service that tracks the registration, unregistration, and modification of services of interest. Given a set of services, a ServiceTracker object begins the tracking by listening to ServiceEvents from the framework, and the actions in the event of service changes can be customized.
XML Parser Service (3)	Defines how XML, SAX, and DOM parsers can be provided and used in an OSGi service platform.
Position (3)	Handles geographical positions and movements in OSGi applications. Based on WGS-84 GPS code, a Position object contains latitude, longitude, altitude, track, and speed fields.
Measurement and State (3)	Allows a consistent handling and exchange of a wide variety of measurements. Any measurement can be represented by the seven basic SI units and derived units. A State object holds integer values to represent discrete states.
Execution Environment (3)	Defines two execution environments for OSGi Server Platform Servers. One is a minimal environment that supports OSGi Framework and implementations of basic services. The other is derived from Foundation Profile. Foundation Profile is a set of Java APIs and provides a complete Java ME application environment for consumer products and embedded devices.

CHAPTER 7

Universal Interactions*

A critical challenge facing the pervasive computing research community is the need to manage complex interactions between the user and the space and among the numerous interconnected computers and devices. In such a pervasive space, a given application's functionalities are partitioned and distributed across several computing devices that are spontaneously discovered and used. In particular, because various devices will need to use the application's user interface, the interface must support and be able to adapt to various interaction modalities, device capabilities, and local computing resources.

In recent years, researchers have devoted much attention to universal interactions with diverse devices in richly networked settings. We can broadly categorize the numerous approaches explored into two groups: (1) universal user interface languages and (2) user interface remoting. Using the universal UI languages approach, developers write the user interaction in an abstract language without targeting any particular device, so it can later be instantiated and presented to the specific device used in the interaction. More specifically, a universal language describes user interfaces that are rendered by mapping a description's device-independent interaction elements to a target platform's concrete interface objects. The UI remoting approach stems from service discovery frameworks (Chapter 5) and enables device interoperability using an agreed-upon user interface presentation protocol to remote devices.

Here, we review recent noteworthy efforts for universal interactions using these two approaches. Such efforts aim to raise interoperability in interactive smart spaces by standardizing user interface languages or communication protocols.

7.1   LANDSCAPE OF UNIVERSAL INTERACTION STANDARDS

R&D efforts to facilitate networked device interactions had its roots in Web accessibility, abstract user interface description, and dynamic service discovery. The proliferation of diverse devices in recent years, together with the explosive adoption of Web portals, has created a need for increased Web accessibility—from any device, by anyone, at any context.

*This chapter is based on the following contribution: Choonhwa Lee, Sumi Helal and Wonjun Lee, "Universal Interactions with Smart Spaces," the Standards, Tools and Emerging Technologies Department, IEEE Pervasive Computing magazine, vol. 5, no. 1, January–March 2006.

Device Independence (DI), an integral part of the World Wide Web Consortium's (W3C) efforts for unconstrained Web access, defines an integrative framework of delivery context and adaptation. The delivery context conveys a client device's characteristics, and a server adapts the pages being accessed, modifying their layout and style to cater to the client's device capability and personal preference. The integrative framework lets us employ constituent component technologies.

For instance, the W3C Composite Capabilities/Preferences Profile (CC/PP) [33] is the first attempt to define a vocabulary to describe delivery context. Also, the W3C has standardized its XForms—a platform-independent markup language for the next generation Web form for achieving device-independent presentations.

In addition to DI efforts, other numerous user interface languages exist, including:

- the International Committee for Information Technology Standards Universal Remote Console (INCITS/V2 URC) [29–31],
- the User Interface Markup Language (UIML) [34],
- the Extensible Interface Markup Language (XIML) [35], and
- Carnegie Mellon University's Personal Universal Controller (PUC) [36].

These languages enable device-independent presentation by letting target devices determine the most suitable presentation from a given universal description in terms of a predefined set of abstract user interface components. The targets might be common IT devices, assistive-technology devices for the physically challenged, or other resource-constrained devices, which require support for various modalities such as visual, auditory, and tactile interfaces. For example, we should be able to present a text component in an abstract description as displayed text on computer screens, spoken audio for the blind, or output on Braille strips.

UI remoting takes a different approach from the universal UI languages. It uses a remote user interface protocol that lets an application interact with its user interface proxy exported to a remote device. The protocol relays I/O events between an application and its user interface, which resides on a remote machine. In this approach, a broader range of devices can control the application—even resource-poor devices that can barely afford the remote user interface protocol.

UI remoting makes even more sense when used within a dynamic-service discovery framework. This is because lookup and matchmaking can be used to locate information about remote user interface protocols supported by applications and client devices and their capabilities. The Universal Plug and Play (UPnP) Remote User Interface (RUI) [22] standard and the μ Jini Proxy architecture (discussed later) [37] belong to this category.

Various approaches to the universal interaction problem can also be classified into different authoring styles, depending on whether the user interface targets specific client device platforms.

The W3C DI articulates three possible cases: single, flexible, or multiple authoring. Single authoring automatically adapts a single generic description to different device capabilities (one size fits all). In contrast, multiple authoring develops a user interface for each type of client devices (custom made). This might offer the most complete user interface and best user experience but at an almost prohibitive development cost. Flexible authoring with a limited set of special user interfaces represents a compromise: customized user interfaces for popular platforms and automatically generated interfaces for rare platforms.

7.2   UNIVERSAL UI LANGUAGES

The two front-runner universal UI languages are W3C XForms and INCITS/V2 URC. However, we also briefly compare UIML, XIML, and CMU PUC.

7.2.1   W3C XForms

W3C XForms differs from HTML forms in that it separates content from presentation [32]. XForms documents consist of two sections: a data model (the XForms model) and data presentation (the XForms user interface and other presentation options). The XForms model is a template of an XML data instance being collected, and the data presentation describes how to display the data. An XForms user interface comprises a predefined set of generic interface elements, called XForms form controls, which capture a high-level logic of user interactions. Each maps to concrete interaction components on target access devices.

Figure 7.1 shows a sample e-commerce form in XForms: an XForms model, its abstract user interface description, and a possible presentation on target devices. The <xforms:instance> element defines an XML template for data to be collected, while <xforms:submission> submits the collected data to the server. XForms form control elements are shown in bold in the abstract description. The <label> element might be displayed on PDAs or be spoken for a blind user. The <select1> element might be rendered as a radio button or spoken options, and the <input> element inputs data. The example also shows that the form controls are bound to XML elements in their corresponding XForms model using the ref binding mechanism. For example, the <select1> element's ref attribute points to the <method> element in the XForms model, defining a link between the two sections of the data model and user interface.

7.2.2   INCITS/V2 Universal Remote Console (URC) Standards

INCITS/V2 URC is a set of standards enabling remote and alternative interfaces for information and electronic products [29–31]. The standards define a generic framework and an XML-based user interface language to let a wide variety of devices act as a remote to control other devices called

```
<xforms:model>
  <xforms:instance>
    <ecommerce xmlns="">
      <method/>
      <number/>
      <expiry/>
    </ecommerce>
  </xforms:instance>
  <xforms:submission action=...
   id="submit" method="post"/>
</xforms:model>
              (a)


<select1 ref="method">
  <label>Select Payment Method:</label>
  <item>
    <label>Cash</label>
    <value>cash</value>
  </item>
  <item>
    <label>Credit</label>
    <value>cc</value>
  </item>
</select1>
<input ref="number">
  <label>Credit Card Number:</label>
</input>
<input ref="expiry">
  <label>Expiration Date:</label>
</input>
<submit submission="submit">
  <label>Submit</label>
</submit>
              (b)
```

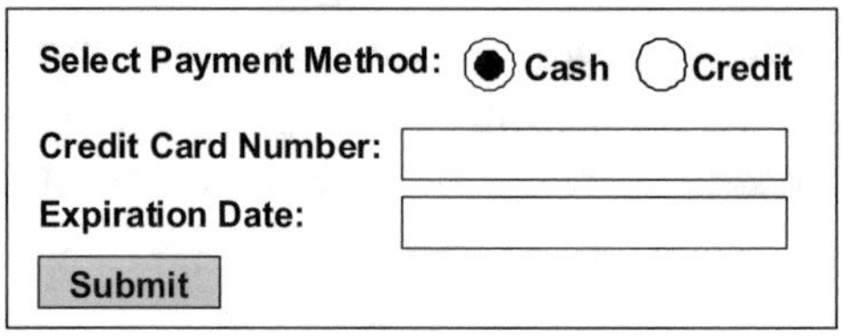

Select Payment Method: ⦿ Cash ◯ Credit

Credit Card Number: [   ]

Expiration Date: [   ]

[Submit]

(c)

FIGURE 7.1: A sample W3C XForms user interface: (a) a sample e-commerce XForms model, (b) its abstract user interface description, and (c) a possible user interface for target devices.

Targets. The URC approach models each Target's functional units as *User Interface Sockets*, which act as an access and control point to the Target. Consider a digital thermometer Target that displays the current temperature, recent minimum and maximum temperatures, and the scale either in Fahrenheit or centigrade [31]. It also lets users reset the minimum and maximum temperature to the current temperature. A UI Socket description in Figure 7.2(a) describes the Target's state and functionality using Variables, Commands, and Notifications. The Variables are state variables to indicate dynamically changing information of the Target—the current temperature. URC invokes the Commands to ask the Target to perform certain functions such as a "reset." The Target triggers Notifications to notify URC users of certain events. In figure 7.2(b), *checkReset* checks on a reset request with its users.

A modality-independent user interface specification—the presentation template in Figure 7.2—is accompanied with the abstract-functionality description. It provides the socket presentation information by describing a structure of abstract interactors, each of which binds to the socket elements. This is much like mapping XForms form controls to a data model element. Moreover, the interactors are largely a subset of XForms 1.0 form controls, with a few exceptions. The presentation template includes the *output* interactor for displaying a value of variables and commands, *select1* for a single choice, *trigger* for triggering a command bound to a command element in the socket description, and *modalDialog* for a target-triggered dialog (which is bound to a *notify* element in the socket description). Note that the ref attribute in each interactor establishes a binding to relevant socket elements. Besides, the *group* element organizes individual interactors in a hierarchical fashion. Figure 7.2(c) shows a possible user interface resulting from the socket description and presentation template.

7.2.3   UIML, XIML, and PUC

Sharing the same philosophy of separating an abstract interface description from its later rendering in any delivery context, UIML, XIML, and CMU's PUC define a set of basic interaction elements, a mechanism for grouping such elements, and optional additional presentation information.

A UIML document [34] has styling sections that map interface elements to target UI objects (such as GUI widget classes), implying the need for one style section for each target device type. However, unlike other user interface languages, UIML doesn't support an explicitly separate data model, resulting in undesirable data fusion in the user interface elements.

An XIML presentation component [35] defines concrete interaction elements for a particular target platform. Similar to UIML, it can support a new device type by defining one presentation component, meaning it follows the multiple authoring approach. Alternatively, a single intermediate presentation component can describe XIML interfaces and automatically create a concrete

```
<uiSocket about="http://www.mycorp.com/thermometer/socket"
    id="socket" xmlns="http://www.incits.org/incits390-2005"
    xmlns:xsd="http://www.w3.org/2001/XMLSchema">
  <variable id="temperature" type="xsd:double">
    <dependency write="false()"/>
  </variable>
  <variable id="maximum" type="xsd:double">
    <dependency write="false()"/>
  </variable>
  <variable id="minimum" type="xsd:double">
    <dependency write="false()"/>
  </variable>
  <variable id="scale" type="scaleType"/>
  <command id="reset"/>
  <notify id="checkReset" explicitAck="false" category="alert">
    <dependency acknowledge="value('confirmReset')='done' or value('cancelReset')='done'"/>
  </notify>
  <command id="confirmReset" type="uiSocket:basicCommand">
    <dependency read="value('checkReset')='active'" execute="value('checkReset')='active'"/>
  </command>
  <command id="cancelReset" type="uiSocket:basicCommand">
    <dependency read="value('checkReset')='active'" execute="value('checkReset')='active'"/>
  </command>
  <xsd:schema>
    <xsd:simpleType name="scaletype" id="idScaleType">
      <xsd:restriction base="xsd:string">
        <xsd:enumeration value="F"/>
        <xsd:enumeration value="C"/>
      </xsd:restriction>
    </xsd:simpleType>
  </xsd:schema>
</uiSocket>
```

(a)

```
<pret name="http://www.mycorp.com/thermometer/corepret"
  id="pret"
    xmlns="http://www.incits.org/incits391-2005">
  <dcterms:conformsTo>http://www.incits.org/incits391-2005</dcterms:conformsTo>
  <group id="readings">
    <output id="temperature" ref="http://www.mycorp.com/thermometer/socket#temperature"/>
    <output id="maximum" ref="http://www.mycorp.com/thermometer/socket#maximum"/>
    <output id="minimum" ref="http://www.mycorp.com/thermometer/socket#minimum"/>
  </group>
  <select1 id="scale" ref="http://www.mycorp.com/thermometer/socket#scale"/>
  <trigger id="reset" ref="http://www.mycorp.com/thermometer/socket#reset"/>
  <modalDialog id="checkReset" ref="http://www.mycorp.com/thermometer/socket#checkReset">
    <trigger id="confirmReset" ref="http://www.mycorp.com/thermometer/socket#confirmReset"/>
    <trigger id="cancelReset" ref="http://www.mycorp.com/thermometer/socket#cancelReset"/>
  </modalDialog>
</pret>
```

(b)

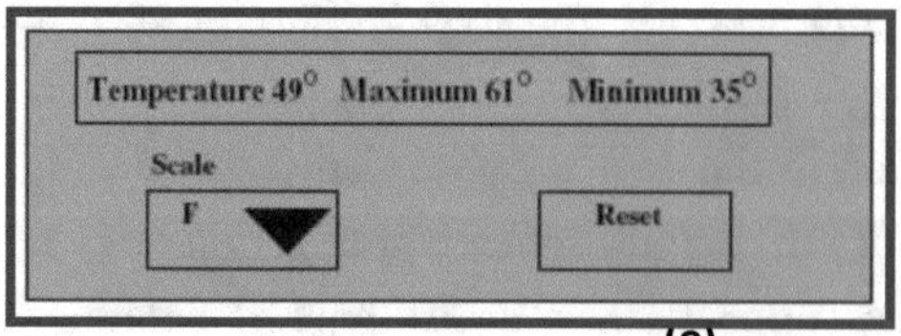

(c)

FIGURE 7.2: A sample INCITS/V2 Universal Remote Console user interface: (a) a digital ther-mometer UI Socket description, (b) the corresponding presentation template, and (c) a possible user interface.

presentation component using a set of predefined relations. Therefore, XIML (like INCITS/V2 URC) belongs to the flexible authoring style that allows multiple implementations, each finetuned to a particular device class, apart from the all-in-one abstract description.

PUC [36] describes device functions in terms of state variables and commands reminiscent of those of INCITS/V2 URC. In addition to its grouping mechanism as a hint for placing relevant components closer together, it can specify dependency information to indicate whether a component is activatable with regard to others. So, it can gray out or remove unusable parts on a small-display device. This feature is somewhat similar to the URC dependency element of Figure 7.2.

7.3   UI REMOTING

With UI remoting, user interfaces reside on a remote platform instead of the one running their applications. This lets a multitude of wired and wireless devices be used as interaction devices. A remote user interface protocol conveys I/O events of key or touch screen inputs and display updates between a user interface and an application.

For example, a home security system residing on a set-top box could rely on various kinds of devices to signal an alarm, provided they speak the same remote user interface protocol. A TV screen in the living room or a touchpad attached to the kitchen refrigerator might display an alarm, depending on a resident's location. The devices could also relay to the security application user responses such as detailed alarm information retrieval and subsequent deactivation.

We could further improve UI remoting's effectiveness by incorporating it into a service discovery framework. The UPnP Remote User Interface (RUI) standard and the μ Jini Proxy Architecture have recently explored such a marriage because it offers more options for solving the user interface presentation problem. First, the device-independent user interface languages we discussed earlier can be instantly used. An abstract service user interface description might be made available to clients as part of service advertisements for a concrete-interface generation, which belongs to the single authoring case. Second, multiple device-specific user interface implementations can be attached to service advertisements, so that clients can select the most appropriate one to their device capability and preference. Therefore, current service discovery frameworks support the flexible authoring style (without additional efforts).

7.3.1   UPnP Remote User Interface

The UPnP RUI standard, first published in September 2004, added UPnP *RUI Client Service* and *RUI Server Service* on top of the basic UPnP device architecture. The standard defines UPnP support, which maps compatible applications and remote UIs and establishes, maintains, and terminates a control session between them. The *RUI Server Service* is a UPnP service that exposes a list of compatible UIs, whereas an RUI Client Service displays the user interfaces on client devices. The matchmaking is based primarily on supported remote I/O (thin-client) protocols on both sides, including AT&T Virtual Network Computing (VNC), the Microsoft Remote Desktop Protocol (RDP), Intel Extended Remote Technology (XRT), and the Internet Engineering Task Force's Lightweight Remote Display Protocol (LRDP).

The UPnP RUI standard supports two usage scenarios. In the first, a UPnP RUI control point connects an RUI client to remote applications. A UPnP RUI server exposes a list of remoteable user interfaces. An RUI server lets an RUI control point on the network browse interfaces by calling the server's *GetCompatibleUIs* action to retrieve a list of interfaces compatible with a target client device. The control point invokes the Connection action on the client, connecting an out-of-band remote I/O to a designated remote-able application. In other words, the connection setup results in a remote UI executing on the client device.

In the second usage model, available user interface lists are directly pushed to a UPnP RUI client, so that it can initiate a connection by itself. A user chooses one of the user interfaces to view and control, using a local user interface selection menu presented on the client device. For this, the RUI client stores the list of compatible user interfaces, which RUI client control points on the network manage using *AddUIListing*, *GetUIListing*, and *RemoveUIListing* action invocations.

7.3.2 The μ Jini Proxy Architecture

Researchers have prototyped and demonstrated a Jini equivalent to the UPnP RUI standard on mobile phones [37]. Jini has been extended into a μ Jini proxy architecture that supports context-aware service discovery as well as UI remoting.

Targeting interactive spaces inhabited by a range of diverse devices, the system functions as a proxy for resource-constrained devices that can't afford to engage in the Jini protocol. The prototype used an AT&T VNC protocol as a remote I/O protocol between client devices and services residing on the proxy side. The Jini service matching mechanism has also been extended to include dynamically changing properties relevant to services and network condition, as well as client devices.

The μ Jini proxy architecture comprises four main components: μ Jini Protocol, Resident Client, Virtual Thin Client (VTC), and μ Jini proxy (Figure 7.3). The Resident Client resides on a Java 2 Micro Edition (J2ME) client device and takes care of interfacing between the client and the proxy server as well as handling remote UI messages through a VNC channel.

The μ Jini proxy performs context-aware service discovery on the network side for the client and executes a dedicated thin-client server for UI-remoting the discovered service. The VTC is

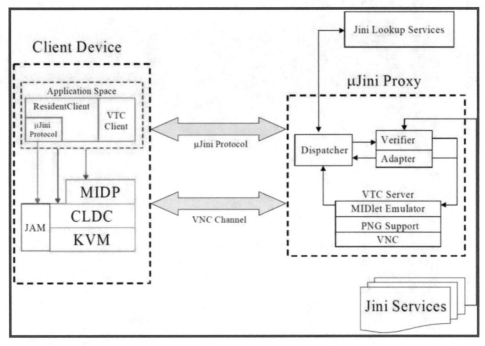

FIGURE 7.3: The μ Jini proxy system architecture.

split into two pieces, VTC Client on the client side and VTC Server on the μ Jini proxy side. The μ Jini protocol lets the client and proxy exchange session management requests and responses, and a separate VNC channel conveys remote I/O events.

It's important to know that the two approaches—universal UI languages and UI remoting—are orthogonal (see Table 7.1 for a comparison of the two approaches). The former doesn't concern itself with service discovery issues, simply assuming an underlying service discovery platform. The latter might be used in parallel with device-independent user interface languages over a service discovery framework, producing synergetic effects.

Furthermore, tackling the same problem naturally leads to overlapping technologies, built on others success. For instance, INCITS/V2 URC seems closer to UPnP than to others in that it first examines the core service information, and then retrieves further details. The two also share the concept of state variables and commands. Another example is that INCITS borrowed most of its URC interactors from W3C XForms. These intertwining and evolutionary standards have been and are pushing the envelope toward a truly universal interaction technology for pervasive spaces.

TABLE 7.1: Comparison of universal UI language and UI remoting.

	UNIVERSAL UI LANGUAGE APPROACH		UI REMOTING APPROACH	
	W3C FORMS	INCITS/V2 URC	UPnP REMOTE UI	μJini PROXY ARCHITECTURE
Origin	Unified Web access	Universal Remote Console	Spontaneous networking	Spontaneous networking
Functionality modeling	Xforms data model	UI socket	UPnP service	Java service interface
Description form	XML/Xpath (Xforms user interface)	XML/RDF (Presentation Temple)	XML/UPnP	Java class
Client and server	Web browser and Web server	URC and Target	UPnP RUIClientService and RUIServerService	Virtual Think Client and Java service
Discovery	External means	Target description and underlying discovery framework	UPnP RUIClient/Server and UPnP Simple Service Discovery Protocol	Enhanced context-aware service discovery
Delivery	Serialized representation	Underlying networking platform	Remote user interface protocols	Remote user interface protocols
User interface description component	Xforms form control	URC interactors	Remote user interface messages	Remote user interface messages

CHAPTER 8

Domestic Robots for Smart Space Interaction*

In this chapter, we take a brief peek at the possible world of robotic companions—humans' co-inhabitants of smart spaces. In such a world, robots are adopted as butlers in our homes, as baby watchers, as friends, and in general, as life companions. In this world, "Helen" lives with Kevin—a robotic companion. In the morning, Kevin checks Helen's schedule and finds out she has an appointment with a neighbor to take a walk at 8:30 a.m. Kevin recognizes that it takes 30 minutes for her to prepare to go for a walk and wakes her up at 8:00 a.m. After she gets up, Kevin reminds Helen that she has an appointment at 8:30 a.m. While she gets dressed, Kevin lets her know that current weather is very cold and suggests that she wears thermal clothing. When she's about to leave the house, Kevin says "good bye" to her and locks the door as she leaves. Kevin hasn't gained wide acceptance. It's not in every home or even in a tiny fraction of homes. However, domestic robots are an impressive technology that is taking shape with several competing products already hitting the marketplace. It could soon be the "chief appliance" in our homes, or the "new interface technology" for smart spaces.

8.1   EXISTING USER INTERFACES IN SMART SPACES

Finding an appropriate user interface for smart space applications that require interactions isn't an easy task. Many projects have attempted interface technologies including PDAs, smart phones, Tablets, wireless wearable microphones, and microphone arrays. In most cases, the interface posed an impediment to the services the user could receive from the smart space. Imagine a frail, 82-year-old elderly person trying to use a PDA to interact with her medicine reminder or to lock all the doors in her house by pressing buttons. Visual or cognitive impairments might make it very difficult for her to use such a PDA. The tablet is bulky and heavy. The smart phone is similarly cumbersome

*This chapter is based on the following contribution: Duckki Lee, Tatsuya Yamazaki, and Sumi Helal, "Robotic Companions for Smart Space Interactions," the Standards, Tools and Emerging Technologies Department, IEEE Pervasive Computing magazine, vol. 8, no. 2, April–June 2009.

to the PDA. In a recent study at the Gator Tech Smart House at the University of Florida, a wireless, wearable microphone with a simple push-to-talk mechanism was used in a three-day live-in trial. Results were totally unexpected. Although the interface seemed easy to use, it hardly functioned according to expectations. The 77-year-old subject couldn't coordinate push-to-talk while actually talking, resulting in incomplete commands that were impossible for the speech recognition engine to map to any known commands.

A new interface technology is clearly needed. The use of robots as multi-modal interface gadgets has begun to receive attention. Several projects and research labs are experimenting with the effectiveness of using robots as interfaces to smart spaces [38–40].

8.2   ROBOTS AS SMART SPACE BUTLERS

Commercially available home robots are available today. Cleaning robots like the iRobot Roomba (www.irobot.com) and entertainment robots like the Sony AIBO (www.sonyaibo.net/home.htm) are widely used. However, the Roomba and the AIBO are specialized robots focused on specific goals, which are cleaning and entertainment, respectively.

While the Roomba tests the waters in consumer acceptability of the robot-at-home concept, a new breed of robots supporting broad space-interaction capabilities is marking its territory. These robots allow for multi-modal interactions with residents and are able to interact and control home appliances. Many of these robots are connected to the rest of the world, acting as gateways to outside services of interest to their users. In the next sections, we review the most significant R&D developments of such robots.

8.2.1   Phyno

The Japanese National Institute of Information and Communications Technology (NICT) created Phyno (see Figure 8.1) and the Yuvi Zoukei Corporation commercialized it. Phyno is a small

FIGURE 8.1: Phyno. An interactive robot developed by NICT, Japan, small enough to be easily placed and used anywhere in a home.

conversational robot equipped with a camera and microphone so that it can interact with people through voice commands. Phyno can also connect to sensors and control home appliances. Additionally, it provides multiple services such as a TV recommendation service, a recipe finder service, and a reminder service. Phyno can't walk, but it's small enough to be located almost anywhere to interact with people anytime. Typically, several Phyno robots are used and placed in different rooms or areas.

8.2.2   ApriAlpha

ApriAlpha (Figure 8.2; www.toshiba.co.jp/about/press/2005_05/pr2001.htm) is currently under development at Toshiba as a robotic information home appliance. Unlike Phyno, it is mobile and has a visual display. It offers various communication capabilities including wireless LAN and Bluetooth. ApriAlpha can perform various tasks around the home. It controls home appliances, recognizes and synthesizes voices, and provides weather, news, and other useful information. ApriAlpha not only recognizes voices but also distinguishes voices from multiple directions. It uses an Open Robot Controller Architecture (ORCA) which enables easy upgrades of its functions.

8.2.3   PaPeRo

NEC developed the Personal Robot PaPeRo (Figure 8.3) and positioned it as "a partner of human being." PaPeRo features advanced functionalities, including finding and identifying people, speech recognition, spontaneous suggestions, autonomous actions, providing weather and news information, controlling TV remotely, and retail and travel recommendations. In addition, PaPeRo shows

FIGURE 8.2: ApriAlpha. Developed by Toshiba, offers many communication capabilities utilizing wireless LAN and Bluetooth, as well as automatic updates and upgrades.

FIGURE 8.3: PaPeRo. Developed by NEC, PaPeRo features advanced functionalities including spontaneous suggestions and autonomous actions, various facial expressions, and scenario-based actions utilizing RoboStudio.

different reactions and facial expressions depending on how a person approaches it. Furthermore, RoboStudio, NEC's scenario-creation software platform enables PaPeRo to detect and interpret specific events for selection and activation of proper scenarios.

8.2.4   Nuvo

Nuvo (see Figure 8.4; http://nuvo.jp/nuvo_home_e.html) is the first home-use humanoid robot and has been commercialized by ZMP in Japan. The main development concept of the Nuvo was

FIGURE 8.4: Nuvo. Nuvo was the first home-use humanoid robot developed with concept of "always being together." It can be controlled with voice commands, a remote controller, or Internet access using a mobile phone.

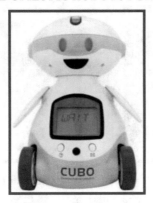

FIGURE 8.5: Cubo. One of three robots developed by INNOmetal IZIrobot in Korea. Mobility is a unique feature that distinguishes Cubo from two other robots (Netoy and Porongbot).

"always be together." In reflection of the concept, Nuvo is small and lightweight for safety and mobility. Nuvo can be controlled with voice commands, a remote controller, or Internet access using a mobile phone.

8.2.5 Cubo, Netoy, and Porongbot

INNOmetal IZIrobotot (www.izirobotics.com/english/products_05.html) in Korea has developed three different robots: Cubo, Netoy, and Porongbot (see Figures 8.5–8.7). As a ubiquitous robot companion (URC), Cubo can provide a variety of services: storytelling, news, weather, music, and home monitoring. Cubo has mobility whereas Netoy and Porongbot don't. Netoy, an emotional network robot, was developed as a personal assistant. It can express emotions through LED and arm movement and provide information services (news and weather), entertainment services (music streaming and MP3 playing), and private secretary services (voice messenger). Porongbot, which is an edutainment network robot, was developed for education and entertainment. Through a

FIGURE 8.6: Netoy. Along with Cubo and Porongbot, Netoy was developed by INNOmetal IZIrobot. Netoy is an emotional networked robot that expresses its emotion through LEDs or arm movements.

FIGURE 8.7: Porongbot. Developed by INNOmetal IZIrobot together with Cubo and Netoy, is an edutainment robot featuring education and entertainment actions with a 3.5-inch touch screen.

3.5-inch touch screen, it provides the following features and functionalities: study action (living play and idea play), entertainment action (children's songs, stories, and games), and occupational action (vocational education and experience). Moreover, a network-based robot software development kit (SDK) enables the efficient creation of network-based robot contents and services.

8.2.6 iRobiQ

Yujin Robot in Korea developed iRobiQ (see Figure 8.8; www.irobibiz.com/english). As a network-based service robot, iRobiQ provides information services for news and weather, entertainment (including showing photos and videos, and even singing karaoke), education services (such as teaching

FIGURE 8.8: IRobiQ. A networked robot develoepd by Yujin Robot in Korea providing diverse everyday services including a remote monitoring service through a mobile phone interface. iRobiQ has a wide 7-inch touch screen that facilitate function control, roams freely and charges itself automatically.

English), and a remote monitoring service through a mobile phone interface. iRobiQ has a 7-inch LCD touch-screen to control its functions. It can roam freely and charge itself automatically when its battery is out. Yujin Robot also features an SDK that includes a robot application interface, script editor, and 3D simulator, which lets programmers and developers easily develop service content. In addition, iRobiQ is based on Windows XP, which means it can convert PC content to robot content with a robot's motion and face expression through a content creator.

8.2.7   Nao

Aldebaran Robotics in France developed "Nao" (see Figure 8.9; http://www.aldebaran-robotics. com/eng/index.php). Designed for entertainment purposes, Nao is able to interact with its owner, with evolving behaviors and functionalities. Additionally, the user should be able to teach Nao new behaviors using a computer with Wi-Fi connectivity. Nao is the most mobile service robot for home use today with 25 degrees of freedom. It is based on a Linux platform and scripted with Urbi, an easy-to-learn programming language, with the option of a graphic interface for beginners or code commands for experts.

FIGURE 8.9: Nao. A networked robot developed by Aldebaran Robot in France primarily for entertainment and highly customized companionship.

TABLE 8.1: Robot features.

	CAMERA	SPEAKER	NETWORK	DISPLAY	MOBILITY	FACIAL EXPRESSION	APPLIANCE CONTROL	LOCAL PROCESSING (FACE, VOICE RECOGNITION)	API / DEVELOP. ENV. SUPPORT	NETWORK SUPPORT
Phyno	✓	✓	✓	✗	✗	✗	✓	Face, Voice	✓	LAN
ApriAlpha	✓	✓	✓	✓	✓	✗	✓	Face, Voice	✓	WLAN Bluetooth
PaPeRo	✓	✓	✓	✗	✓	✓	✓	Face, Voice	✓ Robo-Studio	WLAN
Nuvo	✓	✓	✓	✗	✓	✗	✗	Voice	✗	LAN WLAN
Cubo	✓	✓	✓	✓	✓	✗	✗	✗	✓ Robot SDK	WLAN
Netoy	✗	✓	✓	✓	✗	✓	✗	✗	✓ Robot SDK	WAN
Porongbot	✗	✓	✓	✓	✗	✓	✗	Voice	✓ Robot SDK	WLAN
iRobiQ	✓	✓	✓	✓	✓	✓	✗	Face, Voice	✓ Robot SDK	WLAN
Nao	✓	✓	✓	✓	✓	✓	✗		✓ Robot SDK	LAN WLAN

	SIZE (MM) H*D*W	WEIGHT	COMMERCIAL AVAILABILITY	ENGLISH SUPPORT
Phyno	340*210*260	3 kg	✓ Japan Only	✗
ApriAlpha	380*350*350	9.5 kg	✗	✗
PaPeRo	385*245*248	5.0 kg	✗	✗
Nuvo	390*120*350	2.5 kg	✓	✓
Cubo	230*150*160	1.2 kg	✗	✗
Netoy	200*120*110	0.5 kg	✗	✗
Porongbot	281*173*193	1.4 kg	✗	✗
iRobiQ	450*320*320	7 kg	✓	✓
Nao	580*-*-	4.3 kg	✓	✓

TABLE 8.2: Robot specifications.

8.3   COMPARING DOMESTIC ROBOTS

Table 8.1 compares the robots and lists available or missing features and capabilities. It's interesting to note that all robots are networked (the majority support wireless), all robots are equipped with speakers, most have cameras, and most offer an API and SDK for developers. Table 8.2 compares other aspects including power consumption, dimensions, language support, and commercial availability.

8.4   THE CASE FOR DOMESTIC ROBOTS

We could argue that robotic interfaces are friendlier and more natural interfaces than information appliances such as PDAs, Tablets, and mobile phones. A robot could act like a friend, messenger, entertainer, or teacher, whereas a traditional interface to smart spaces lacks personality. Also, the robot's facial expression capabilities enhance its friendliness and help it to act in specific roles and communicate more accurate responses to users. Robots are arguably more natural to interact with than other interfaces because robots mimic interacting with other people, which is as natural as it could be achieved when compared with using pull down menus or a stylus on a PDA. From a performance perspective, robots can provide better voice interaction because of their ability to accompany the user, allowing for a better and more natural way of filtering noise. Finally, customizing the interaction based on user profiles is handled more easily and naturally through face and voice recognition.

.   .   .   .

CHAPTER 9

Continua:
An Interoperable Personal Health
Echosystem*

Soaring costs and declining quality and availability are pressing governments to reform Healthcare. More scalable and cost-effective approaches are being sought by exploring technological advances in devices, sensors, and interconnectivity. Specifically, to improve the quality and affordability of health care for all, new *point-of-care* delivery methods are being engineered and developed. Various technologies could help by extending treatment and care beyond traditional clinical settings into personal and home settings. However, creating such a personal telehealth ecosystem requires high degree of interoperability. This is a major challenge given that most personal health devices are either not connected, or their connectivity to enterprise services is mainly proprietary.

The Continua Health Alliance [43] was formed in 2006 to establish an ecosystem of connected personal health and fitness products and services, making it possible for patients, caregivers and health care providers to more proactively address ongoing healthcare needs. The Alliance has been creating design guidelines that are based on proven, existing connectivity standards. It has also created a product certification program with a consumer recognizable logo signifying the promise of interoperability with other certified products. Continua has quickly grown to become an international alliance of over 200 companies.

9.1   INTEROPERABILITY THROUGH INDUSTRY STANDARDS
Figure 9.1 gives an example overview of a typical ecosystem of personal telehealth devices and services. Continua aims to enable the alignment of different vendors and domains, focusing on

---

*This chapter is based on the following contribution: Randy Carroll, Rick Cnossen, Mark Schnell, and David Simons, "Continua: An Interoperable Personal Healthcare Ecosystem," the Standards, Tools and Emerging Technologies Department, IEEE Pervasive Computing magazine, vol. 6, no. 4, October–December 2007.

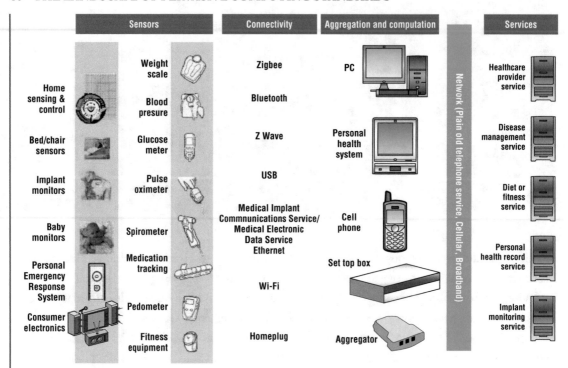

FIGURE 9.1: A typical personal telehealth ecosystem.

- disease management: managing a chronic disease outside of a clinical setting,
- aging independently: using technology and services to live in your own home longer, and
- health and fitness: expanding personal health and wellness to where you live and play.

The process Continua uses to develop its interoperability guidelines is centered on the use of industry standards. It starts by evaluating member-submitted use cases about interoperability problems related to one of the three focus areas. It then collapses the submittals into a consolidated, generalized list of use cases. Continua uses this list to prioritize capabilities, interfaces, and devices and then derives the desired functionality and requirements for the next version of the guidelines.

After this, Continua canvasses the healthcare industry for existing standard development organizations (SDOs) and standards that best satisfy questions such as

- How well do the standards address the capabilities in the selected use cases?
- Does the SDO have international standards or a path to generate them?
- Can Continua member companies participate in the SDO?

- How well is the standard harmonized with related domain standards?
- What are the specification access and control mechanisms?
- What is the associated intellectual-property model?
- Is there tool support?
- What is the level of adoption and maturity?

Once Continua selects the candidate standards, it compares them against the requirements to identify and address any gaps.

Ultimately, the interoperability guidelines defines profiles over the standards and serves as a basis for product certification. A product that passes certification displays the Continua interoperability logo.

9.2   REFERENCE ARCHITECTURE

The Continua end-to-end reference architecture gives a high-level architectural view of the Continua ecosystem, including its topology constraints (Figure 9.2). The distributed-systems architecture breaks down its functionality into five reference-device classes and four network interfaces that connect the devices to a reference topology. The network interfaces are at the center of Continua's interoperability goals and are the crux of the test and certification targets for candidate devices.

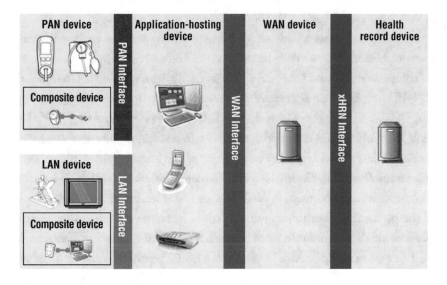

FIGURE 9.2: The Continua end-to-end reference architecture.

The Peripheral Area Network Interface (PAN-IF) connects an application-hosting device, such as a personal computer, cell phone, or monitoring hub, to a PAN device, which is either a sensor or an actuator. A sensor might be a glucose meter, weight scale, pedometer, heart-rate monitor, or carbon monoxide detector. The actuator could be a device that can turn light on or off, shut off the gas in an emergency, output text, or set off an alarm. The PAN-IF has both a lower-layers component (encompassing the classic open-systems interconnection layers 1–4) and an upper-layers component (encompassing the classic OSI layers 5–7).

Example instantiations of the PAN-IF lower layers include both wired and wireless links such as USB- and Bluetooth-based technologies. The PAN-IF upper layers are implemented using the ISO/IEEE 11073-20601 Optimized Exchange Protocol, which leverages work from the ISO/IEEE 11073 Medical Device Communications working group.

The Local Area Network Interface(LAN-IF) connects an application-hosting device to a LAN device. This device aggregates and shares (though a network) the bound PAN devices' information (this is often referred to as a proxy function). A LAN device can also implement sensor and actuator functionality directly. This means that the LAN-IF upper layers can support the same device data model as the PAN-IF upper layers (that is, the ISO/IEEE 11073-20601 data model). Using the same device data model, regardless of the underlying lower-layers communications mechanism, is a key interoperability feature. Continua aims to base the LAN-IF lower layers on Internet Protocol technology to enable different IP-centric communications technologies such as Ethernet and Wi-Fi technologies.

The Wide Area Network Interface (WAN-IF) connects an application-hosting device to one or more WAN devices. A typical WAN device implements a managed-network-based service. It collects information and hosts a wide range of value-adding services such as health- or fitness-monitoring service hosted on a network-based server. The WAN-IF upper layers use a device data model that is compatible with the LAN-IF upper layers' device data model. Continua also plans to base the WAN-IF lower layers on IP technology to enable IP-centric communications technologies, such as xDSL, DOCSIS (Data over Cable Service Interface Specifications), PPP/POTS (Point-to-Point Protocol/ Plain old telephone service), and 3G wireless communication. Again, the sharable, exchangeable device data model is the key component of the interoperable Continua ecosystem.

The Electronic/Personal Health Records Network Interface (xHRN-IF) enables patient-centric data communications between a WAN device and a health-record device, typically at the boundary of the personal telehealth ecosystem. This is in contrast to the other interfaces, which support device-centric data communications between an application hosting device and other Continua devices. The typical xHRN device implements a health record database or other system, managed and operated by a traditional healthcare service provider (e.g., an electronic-health-records system that a hospital or healthcare system manages and operates). The xHRN-IF lets multiple enterprise healthcare entities exchange personal health information. The corresponding health record systems have existing industry standard information models that likely differ from the Continua WAN device. This interface describes how healthcare entities can transform the data so that all parts of the larger healthcare systems can collaborate.

All of these Continua interfaces will have associated guideline specifications and test suites. However, Continua can't encompass all the communications interfaces that various vendors bring to the market using existing or emerging proprietary or open technologies. Hence, noncertified interfaces will exist in the personal telehealth ecosystem that aren't part of the Continua reference architecture. However, the architecture will be able to bridge devices with noncertified interfaces to the Continua ecosystem using a PAN adapter device or a LAN sharing device. For example, an RF-receiver dongle paired via a proprietary wireless communications technology to a health watch may, as a set, be certified as a Continua PAN device. The architecture can then plug that device into Continua application-hosting devices.

9.3   STATUS

To facilitate this large-scale operation, Continua has created a series of working groups, all governed by a board of directors. These groups pursue independent subgoals and tasks and periodically report to the larger Continua organization. The Technical Working Group (TWG) has organized its work into numerous subgroups, including one for each of the four interoperable interfaces and one for each of the three focus areas. It also has subgroups that focus on some overarching subjects, such as the overall architecture or system security and privacy.

In February 2009, Continua announced the completion of its first set of guidelines known as *Version One Design Guidelines*. The guidelines are available for public access and use from the Continua Alliance web site. Additional capabilities are being addressed by the various TWG's subgroups, which will be made available through the release of future versions of the guidelines. In the mean time, through Continua's certification and testing program, many personal medical devices have been certified and made available in the market today.

· · · ·

References

[1]   A. Dey, D. Salber, and G. Abowd, "A Conceptual Framework and a Toolkit for Supporting the Rapid Prototyping of Context-Aware Applications," *Human Computer Interaction J.*, vol. 16, 2001, pp. 97–166. doi:10.1207/S15327051HCI16234_02

[2]   LabView. www.ni.com/labview.

[3]   EchoNet. www.echonet.gr.jp/english/8_kikaku/index.htm.

[4]   Detailed Stipulations for ECHONET 1.0 Device Objects, http://www.echonet.gr.jp/english/spec/pdf/spec_v1e_appendix.pdf.

[5]   IEEE 1451. http://ieee1451.nist.gov.

[6]   SensorML. www.opengeospatial.org/standards/sensorml.

[7]   DeviceKit. www.eclipse.org/ohf/components/soda/index.php.

[8]   Device Description Language (DDL). www.icta.ufl.edu/atalas/programmers_manual.htm.

[9]   C. Chen and A. Helal, "Device Integration in SODA using the Device Description Language," Proceedings of the IEEE/IPSJ Symposium on Applications and the Internet, July 2009, Seattle, Washington, USA.

[10]   L. Nachman, R. Kling, R. Adler, J. Huang, and V. Hummel, "The Intel Mote Platform: A Bluetooth-Based Sensor Network for Industrial Monitoring," 4th Int'l Conf. Information Processing in Sensor Networks, Springer, 2005, pp. 437–442. doi:10.1109/IPSN.2005.1440968

[11]   J. King, R. Bose, H. Yang, S. Pickles, and A. Helal, "Atlas—A Service-Oriented Sensor Platform," Proc. 1st IEEE Int'l Workshop on Practical Issues in Building Sensor Network Applications (SenseApp 2006), IEEE CS Press, 2006.

[12]   S. Karnouskos et al., "Integration of SOA-Ready Networked Embedded Devices in Enterprise Systems via a Cross-Layered Web Service Infrastructure," IEEE Conf. Emerging Technologies and Factory Automation (ETFA 2007), IEEE Press, 2007, pp. 293–300.

[13]   D. Zhang et al., "Handling Heterogeneous Device Interaction in Smart Spaces," Ubiquitous Intelligence and Computing, Springer, 2006, pp. 250–259. doi:10.1007/11833529_26

[14]   SODA. www.eclipse.org/ohf/components/soda/stepstone. php.

[15]   T. Erl, Service-Oriented Architecture: Concepts, Technology, and Design, Prentice Hall, 2005.

[16]   D. Krafzig, K. Banke, and D. Slama, Enterprise SOA: Service-Oriented Architecture Best Practices, Prentice Hall, 2005.

[17]   T. Erl, Service-Oriented Architecture: A Field Guide to Integrating XML and Web Services, Prentice Hall, 2004.

[18]   H. Havenstein, "SOA App Quickly Boosts Storm Response," Computer-World, June 2006; www.computerworld.com/action/article.do?command= viewArt icleBasic& articleId=112207.

[19]   C. Koch, "How SOA Really Works," CIO, Aug. 2005; www.cio.com/blog_ view.html? CID=10591.

[20]   A. Stanford-Clark, "Coupled or Decoupled Plus Heavyweight and Lightweight Delivery Considerations in an Enterprise Service Bus Context," Middlewarespectra, Aug. 2004, pp. 26–33.

[21]   The Jini Service Lookup and Delivery Standard. www.jini.org.

[22]   Universal Plug and Play (UPnP). www.upnp.org.

[23]   Service Location Protocol (SLP). www.ietf.org.

[24]   Bluetooth Service Discovery Protocol. www.bluetooth.com.

[25]   D. Marples and P. Kriens, "The Open Services Gateway Initiative: An Introductory Review," IEEE Communications Magazine, vol. 39, no. 12, Dec. 2001, pp. 110–114.

[26]   G. Bell, "Bell's Law for the Birth and Death of Computer Classes," Communications of the ACM, January 2008, Vol. 51, No. 1, pp. 86–94. doi:10.1145/1327452.1327453

[27]   Crossbow. http://www.xbow.com.

[28]   S. Greenberg and C. Fitchett, "Phidgets: Easy Development of Physical Interfaces through Physical Widgets," Proc. 14th Annual ACM Symposium on User Interface Software and Technology (UIST 01), ACM Press, 2001, pp. 209–218.

[29]   The V2 Universal Remote Console Standard. http://www.incits.org/tc_home/v2.htm.

[30]   G. Vanderheiden and G. Zimmermann, Interface Sockets, Remote Consoles, and Natural Language Agents—A V2 URC Standards Whitepaper. (myurc.org/whitepaper.php).

[31]   INCITS/V2, ANSI INCITS 389-2005: "Protocol to Facilitate Operation of Information and Electronic Products through Remote and Alternative Interfaces and Intelligent Agents: Universal Remote Console," Feb. 2005, www.ncits.org/list_INCITS.htm.

[32]   M. Dubinko et al., XForms 1.0, World Wide Web Consortium (W3C) recommendation, Oct. 2003; www.w3.org/TR/2003/REC-xforms-20031014.

[33]   Composite Capabilities/Preference Profiles: Structure and Vocabularies (CC/PP). The Ubiquitous Web Applications Working Group (UWAWG). www.w3.org/Mobile/CCPP/.

[34]   User Interface Markup Language (UMIL). http://uiml.org/.

[35]   Extensible Interface Markup Language (XIML). http://www.ximl.org/.

[36]   The Pittsburgh Pebbles PDA Project. http://www.pebbles.hcii.cmu.edu/.

[37]   C. Lee, A. Helal, and D. Nordstedt, "μ Jini Proxy Architecture for Impromptu Mobile Service Access," *Proc. Workshop Next Generation Service Platforms for Future Mobile Systems* (SPMS 06), 2006.

[38]   T. Yamazaki and T. Toyomura, "Sharing of Real-Life Experimental Data in Smart Home and Data Analysis Tool Development," Proc. 6th Int'l Conf. Smart Homes and Health Telematics (ICOST 08), Springer, 2008, pp. 161–168. doi:10.1007/978-3-540-69916-3_19

[39]   N. Matsumoto et al., "An Intelligent Artifact as a Cohabitant: An Analysis of a Home Robot's Conversation Log," Proc. 2nd Innovative Computing, Information and Control (ICICIC 07), IEEE CS Press, 2007, pp. 21–21.

[40]   A. Helal et al., "Experience of Enhancing the Space Sensing of Networked Robots Using Atlas Service-Oriented Architecture," Proc. of the 8th Asia-Pacific Conference on Computer Human Interaction (APCHI 08), Springer, 2008, pp. 1–10. doi:10.1007/978-3-540-70585-7_1

[41]   Sun SPOT Official Web Site. www.sunspotworld.com.

[42]   L. Holmquist et al., "Smart-Its Friends: A Technique for Users to Easily Establish Connections between Smart Artefacts," Proc. 3rd Int'l Conf. Ubiquitous Computing (UBICOMP 01), Springer-Verlag, 2001, pp. 116–122. doi:10.1007/3-540-45427-6_10

[43]   The Continua Health Alliance. www.continuaalliance.org.

Author Biography

Sumi Helal is a Professor at the Computer and Information Science and Engineering Department (CISE) at the University of Florida. His research interests span the areas of Pervasive Computing, Mobile Computing, and networking and Internet Computing. He directs the Mobile and Pervasive Computing Laboratory at the CISE department, and is co-founder and director of the Gator Tech Smart House—an experimental home for applied research in the domain of aging, disability, and independence. He led the technology development of the NIDRR-funded Rehabilitation Engineering Research Center (RERC) on Successful Aging (2001–2007) and is currently leading a new initiative on smart home based personal health and independence, funded by the National Institutes of Health.

Outside of his teaching and research activities, Dr. Helal is the Associate Editor in Chief of IEEE Computer, an editorial board member of the IEEE Pervasive Computing magazine and Editor of its column on Standards, Tools and Emerging Technologies. He has been on the editorial board of the IEEE Transaction on Mobile Computing. He published over 200 books, book chapters, journal articles, and conference or workshop papers. He is a senior member of the Institute of Electrical and Electronics Engineers (IEEE), and a member of the Association for Computing Machinery (ACM) and the USENIX Association.

Printed in the United States
by Baker & Taylor Publisher Services